XML Retrieval

Synthesis Lectures on Information Concepts, Retrieval, and Services

Editor
Gary Marchionini, ***University of North Carolina, Chapel Hill***

XML Retrieval
Mounia Lalmas
2009

Understanding User–Web Interactions via Web Analysis
Bernard J. (Jim) Jansen
2009

Faceted Search
Daniel Tunkelang
2009

Introduction to Webometrics: Quantitative Web Research for the Social Sciences
Michael Thelwall
2009

Exploratory Search: Beyond the Query-Response Paradigm
Ryen W. White, Resa A. Roth
2009

New Concepts in Digital Reference
R. David Lankes
2009

Automated Metadata in Multimedia Information Systems: Creation, Refinement, Use in Surrogates, and Evaluation
Michael G. Christel
2009

XML Retrieval

Mounia Lalmas

www.morganclaypool.com

ISBN: 9781598297867 paperback
ISBN: 9781598297874 ebook

DOI 10.2200/S00203ED1V01Y200907ICR007

A Publication in the Morgan & Claypool Publishers series
SYNTHESIS LECTURES ON INFORMATION CONCEPTS, RETRIEVAL, AND SERVICES

Lecture #7
Series Editor: Gary Marchionini, *University of North Carolina, Chapel Hill*

Series ISSN
Synthesis Lectures on Information Concepts, Retrieval, and Services
Print 1947-945X Electronic 1947-9468

XML Retrieval

Mounia Lalmas
University of Glasgow

SYNTHESIS LECTURES ON INFORMATION CONCEPTS, RETRIEVAL, AND SERVICES #7

ABSTRACT

Documents usually have a content and a structure. The content refers to the text of the document, whereas the structure refers to how a document is logically organized. An increasingly common way to encode the structure is through the use of a mark-up language. Nowadays, the most widely used mark-up language for representing structure is the eXtensible Mark-up Language (XML). XML can be used to provide a focused access to documents, i.e. returning XML elements, such as sections and paragraphs, instead of whole documents in response to a query. Such focused strategies are of particular benefit for information repositories containing long documents, or documents covering a wide variety of topics, where users are directed to the most relevant content within a document. The increased adoption of XML to represent a document structure requires the development of tools to effectively access documents marked-up in XML. This book provides a detailed description of query languages, indexing strategies, ranking algorithms, presentation scenarios developed to access XML documents. Major advances in XML retrival were seen from 2002 as a result of INEX, the Initiative for Evaluation of XML Retrieval. INEX, also described in this book, provided test sets for evaluating XML retrieval effectiveness. Many of the developments and results described in this book were investigated within INEX.

KEYWORDS

element, structure, structured document retrieval, focused retrieval, element retrieval, XML, query languages, indexing strategies, ranking algorithms, presentation scenarios, INEX

Contents

Acknowledgments

This book is based on a book chapter on Structured Text Retrieval to appear in the second edition of [15], an entry on XML Retrieval to appear in the Encyclopedia of Database Systems [137], and an entry on XML Information Retrieval to appear in the Encyclopedia of Library and Information Sciences [17]. This book and these three articles were written in parallel to entries written for the Encyclopedia of Database Systems [137] in the area of structured text retrieval. Many of these entries influenced the content of this book, and vice versa. These entries include "Structured Document Retrieval" by Lalmas and Baeza-Yates [107] (Chapter 3), "Structured Text Retrieval Models" by Hiemstra and Baeza-Yates [71] (Section 3.3), "Narrowed Extended XPath I (NEXI)" by Trotman [179] (Section 4.3.2), "Indexing Units" by Kamps [84] (Chapter 5), "Contextualization" by Kekalaineen et al [100] (Section 6.2), "Propagation-based structured text retrieval" by Pinel-Sauvagnat [141] (Section 6.3), "Aggregation-based Structured Text Retrieval" by Tsikrika [187] (Section 6.4), "Processing structural constraints" by Trotman [180] (Section 6.6), "Presenting structured text retrieval results" by Kamps [85] (Chapter 7), "Processing overlaps" by Ramirez ([150] (Section 7.1), and "INitiative for the Evaluation of XML retrieval (INEX)" by Kazai [89] (Chapter 8). Earlier versions of Chapter 8 appeared in [113, 114].

A great thank you to Elham Ashoori, and in particular to Benjamin Piwowarski (who went through several rounds of this book) for their comments and feedbacks, and to Gabriella Kazai as this book borrows content from INEX papers we published together and her PhD thesis. I am grateful to Norbert Fuhr and Ian Ruthven for their positive and constructive reviews. I am, however, responsible for all mistakes, omissions and typos.

I would like to acknowledge the great contribution and enthusiasm of people around the world to INEX. They are the ones that have advanced state-of-the-art research in XML retrieval.

Finally, but not least, this book was mostly written during evenings (often accompanied with a nice glass of wine), or while sitting looking at my son in his various sport activities. I am thankful to Thomas Roelleke, my partner in life, and our son for their great patience. This book is dedicated to them.

Mounia Lalmas
June 2009

CHAPTER 1

Introduction

Documents usually have a content and a structure. The content refers to the text of the document, that is words put together to form sentences. The structure refers to how a document is logically organized. For example, a scientific article is usually composed of a title, an abstract, several sections, each composed of paragraphs. An increasingly common way to encode the structure of a document is through the use of a mark-up language. Nowadays, the most widely used mark-up language for representing the structure of a document is the eXtensible Mark-up Language (XML)[1], which is the W3C (World Wide Web Consortium[2]) adopted standard for information repositories and exchanges.

The XML mark-up can be used to provide a focused access to documents, as it can be exploited by so-called XML retrieval systems to return document components, e.g., sections, instead of whole documents in response to a user query. Such focused strategies are believed to be of particular benefit for information repositories containing long documents, or documents covering a wide variety of topics (e.g. books, user manuals, legal documents), where the user's effort to locate relevant content within a document can be reduced by directing them to the most relevant document components, i.e. the most relevant XML elements. In this book, we use document component, XML element or simply element, interchangeably.

The increased adoption of XML as the standard for representing a document structure requires the development of tools to effectively access documents marked-up in XML, from now on referred to as XML documents. The development of such tools has generated a wealth of issues that are being researched by the XML retrieval community. These issues can be classified into five main themes, which correspond to the main stages of an information retrieval process, namely, querying, indexing, ranking, presenting and evaluating:

- **Querying**: A query language may be used at the querying stage to formulate requests with constraints on the content and the structure of the document components to be retrieved. This book will describe query languages that were developed by the W3C (e.g. XQuery Full-Text) and the XML retrieval (e.g. NEXI) communities.

- **Indexing**: XML can be used at the indexing stage, where document components are identified and indexed as separate, but related, units. This book will describe indexing strategies that were developed for XML retrieval.

[1] http://www.w3.org/XML/
[2] http://www.w3.org/

- **Ranking**: XML can be used at the retrieval stage by allowing document components of varying granularity to be retrieved. This book will describe ranking strategies that were developed to return the most relevant document components, ranked in order of estimated relevance.

- **Presenting**: XML can be used at the result presentation stage to organize retrieved document components in the interface in a way that is more useful to users than a simple ranked list of document components. This book will discuss presentation strategies that were proposed to cater for different user scenarios.

- **Evaluating**: The evaluation of XML retrieval requires test collections and measures where the evaluation paradigms are provided according to criteria that take into account the structural aspects of XML documents. This book will describe the significant changes from standard information retrieval evaluation that were necessary to properly evaluate XML retrieval.

Research in XML retrieval has been going on since the late 1990s; however, major advances were seen from 2002 as a result of the set-up of INEX, the Initiative for the Evaluation of XML Retrieval [47, 53, 55, 54, 56, 52, 60], which provided test sets and a forum for the evaluation and comparison of XML retrieval approaches. Many of the developments and results on XML retrieval described in this book were investigated within the INEX evaluation campaigns.

This book has eight parts. It starts by introducing the basics concepts of XML (Chapter 2). Only XML concepts necessary to understand this book are introduced. It then continues with some historical perspectives (Chapter 3), i.e. areas of research that have influenced XML retrieval. The five main parts are, as listed above, query languages (Chapter 4), indexing strategies (Chapter 5), ranking strategies (Chapter 6), presentation strategies (Chapter 7) and evaluation (Chapter 8) for XML retrieval. This book finishes with some conclusions (Chapter 9).

CHAPTER 2

Basic XML Concepts

This chapter describes the basics concepts of XML. The description is by no mean intended to be exhaustive, as here we focus on the concepts that are of concern to XML retrieval, as reviewed in this book. Full details about XML, its design, specifications, tools and applications, can be found in the W3C Recommendation document, published on November 26, 2008[1].

XML, which stands for eXtensible Markup Language, is a simplified version of SGML, which itself stands for Standard Generalized Markup Language. SGML is a meta-language[2] that can be used for example by developers to define markup languages as a means to provide an explicit interpretation of texts independently of devices and systems. SGML is more flexible and powerful than XML but at the expense of being more complex and expensive to implement. Nonetheless, XML retains the key SGML advantages of extensibility, structure and validation but is designed to be simpler and easier to learn and use than SGML.

XML is comparable to HTML (Hyper Text Markup Language) but unlike HTML it allows users to create their own tags. XML tags are self-describing means to encode text so that it can be processed with relatively little human intervention and exchanged across diverse hardware, operating systems, and applications. XML is extensible, in the sense that tags can be defined by individuals or organizations for specific applications, whereas HTML tags have been defined by the W3C.

In HTML, the role of a tag is to instruct the web browser how a particular text fragment should be displayed. In XML, the tags generally define the structure and content of the text, where the actual appearance is specified by a separate application or an associated stylesheet. XML follows the fundamental concept of separating the logical structure of a document from its layout, where the latter is usually described in separate stylesheets. It is the exploitation of the logical structure that is of primary interest in XML retrieval.

It should be said that SGML also allows the representation of a document logical structure. However, SGML has not been widely adopted due to its inherent complexity, apart in specialized areas such as publishing. XML attempts to provide a similar function to SGML, but is less complex.

The basic XML concepts described in this chapter are element (Section 2.1), well-formed XML (Section 2.2), DTD (Section 2.3) and XML Schema (Section 2.4). The chapter ends by showing how XML documents can be viewed as trees (Section 2.5).

[1] Extensible Markup Language (XML) 1.0 (Fifth Edition), W3C Recommendation 26 November 2008, http://www.w3.org/TR/xml/.
[2] http://www.w3.org/MarkUp/SGML/

2.1 ELEMENT

To define a document logical structure, i.e. its logical components, XML adds tags to the document text. The logical components of a document are called XML elements. Each XML element is surrounded by a begin tag and an end tag where the tag denotes the element type and the tokens < , >, </, and /> are the tag delimiters.

Figure 2.1 shows an example of an XML document. An opening tag is `<book>` and its corresponding closing tag is `</book>`. The first line of this document is the processing instruction declaring this document as being in XML format. It also defines the XML specification version. In our example, the document conforms to version 1.0.

The document has one root element, here `<library>`. Each element, including the root element, may contain one or more elements, the boundaries of which are delimited by begin and end tags. For instance, `<book>` is an element with four child elements, also referred to as sub-elements, namely `<title>`, `<author>`, `<year>` and `<abstract>`. `<book>` is said to be the parent element of these four sub-elements. `<library>` has as sub-elements several elements of the same type, namely `<book>`.

The text of a document, i.e. its actual content, is contained within the elements.

Elements can be empty to denote that they do not contain any content, whether text or sub-elements. This can be denoted using the empty element tag, e.g. `<abstract/>` in our example.

Elements can have attributes. In our example, the `<title>` tag has an attribute named "isbn". Attributes are used to associate name-value pairs with elements. Attributes are placed inside begin tags and the attribute value is enclosed in quotes. A given attribute name may only occur once within a tag, whereas several elements with the same tag may be contained within an element, as is the case in our example of `<book>` elements.

In content-oriented documents, attributes are part of the markup, whereas element contents are part of the document contents. In data-oriented documents, the difference is not so clear, for example:

```
<account account-nb="1234"> ... </account>
<account> <account-nb>1234</account-nb> ... </account>
```

The general convention is that attributes should be mostly used as identifiers of elements, whereas the elements should bear the actual document contents.

It is the effective retrieval of XML elements that XML retrieval is concerned with.

2.2 WELL-FORMED XML DOCUMENT

Every XML document must be well-formed before it can be passed to any application, including for the purpose of displaying it. An XML document is said to be well-formed if it satisfies the following rules:

- An XML document should start with `<?xml version "1.0">`, specifying that it is an XML document, and the version of XML, which is currently "1.0".

```
<?xml version="1.0" ?>
<library>
  <book>
    <title isbn="1234">XML Retrieval</title>
    <author>Mounia Lalmas</author>
    <year>2008</year>
    <abstract>This book describes the latest in the
    area of XML retrieval. It discusses work presented
    at INEX, where ....
     </abstract>
  </book>
  <book>
    <title isbn="2468">Web Mining</title>
    <author>Ricardo Baeza-Yates</author>
    <year>2007</year>
    <abstract/>
  </book>
  <book>
    .
    .
  </book>
  .
  .
  .
</library>
```

Figure 2.1: Example of an XML Document

- All XML elements must have a closing tag. If an element has no content, the empty tag can be used. For example, `<title>` is not well-formed, whereas `<title>XML retrieval</title>`, `<title/>` and `<title></title>` are.

- All XML elements must be properly nested within each other. For instance, `<a><b>this is some text</b></a>` is well-formed, whereas `<a><b>this is some text</a></b>` is not. In other words, the begin tag and the end tag must appear in the content of the same element, i. e., elements must nest properly within each other and cannot overlap partially.

- There is one element, referred to as the root element, that encloses the entire document. In our example, this is the `<library>` element.

```
<!DOCTYPE library [
      <!ELEMENT library (book*)>
      <!ELEMENT book (title, author, year, abstract?)>
      <!ELEMENT title (#PCDATA)>
      <!ATTLIST title isbn CDATA "0">
      <!ELEMENT author (#PCDATA)>
      <!ELEMENT year (#PCDATA)>
      <!ELEMENT abstract (#PCDATA)>
]>
```

Figure 2.2: DTD for our example XML document

- Attribute values must always be quoted, either in single or double quotes. However, double quote should be used if the value of the attribute contains single quote marks, e.g. `<author name = "Keith 'C.J.' van Rijsbergen">`, whereas single quote should be used if the value of the attribute contains double quote marks, e.g. `<author name = 'Keith "C.J." van Rijsbergen'>`.
- Tag and attribute names are case-sensitive. Thus, for instance, `<firstname>`, `<Firstname>` and `<FirstName>` are different tags.

2.3 DOCUMENT TYPE DECLARATION

In a given application, XML documents are often requited to be defined according to a specific vocabulary (names of elements (tags) and that of their attributes) and a syntax (grammar defining the document logical structure). Document Type Definition, referred to as DTD, are used to provide this definition.

A DTD lists the element names that can occur in a document, which elements can appear in combination with others, how elements can be nested, what attributes are available for each element type, and so on. The DTD is specified using the EBNF notation[3], and not XML, as is the case for XML Schema (described in Section 2.4).

As an illustration, Figure 2.2 show the DTD for our example document. For instance, `<!ELEMENT book (title, author, year, abstract?)>` states that the `<book>` element is made of three sub-elements, namely `<title>`, `<author>` and `<year>` and sometimes a fourth sub-element `<abstract>`. Further, `<!ELEMENT title (#PCDATA)>` states that `<title>` has no sub-element. PCDATA is a basic type, called parsable character data, and means that the data contained, for instance, between the begin and end `title` tags is text (which can be empty).

Regarding the specification of attributes, `<!ATTLIST title isbn CDATA "0">` expresses that the `<title>` element has an attribute, "isbn", and the attribute type is CDATA (in other words,

[3]http://en.wikipedia.org/wiki/Extended_Backus-Naur _form

is character data) and its default value is "0". CDATA is also a basic type, which represents character data.

Finally, it is possible to specify that some elements may appear several times and some may not appear. In the DTD, this is specified using + for one or more times, * for zero or more times, and ? for an optional element, that is one or not at all. For instance, `<!ELEMENT library (book*)>` specifies that the element `<library>` can have none to several `<book>` elements as children. `abstract?` means that each book may have an abstract.

The specification of a DTD is optional, but is required if the application needs to validate the XML documents. An XML document that conforms to its DTD, when given, is said to be valid.

Having access to a DTD can be useful in XML retrieval, for example, to develop approaches where the semantics associated with the type of an element can be exploited for querying, indexing, ranking or presentation purposes.

2.4 XML SCHEMA

DTDs have limitations. First, they are written in a non-XML syntax, which means that separate tools are needed to process the DTD. Second, they offer very limited data typing[4], i.e. `PCDATA` or `CDATA`, which means string. Third, they have no support for namespaces. The latter are necessary to avoid name collisions for elements and attributes that have the same name but defined in different vocabularies. This happens when tag and attribute names from multiple sources are mixed, as is the case with semantic web applications and distributed digital libraries.

XML Schema provides a more comprehensive method of specifying the structure of an XML document. The schema is itself an XML document, and so can be processed by the same tools that process the XML documents it specifies. Namespaces are declared using "xmlns", the value of which is a Uniform Resource Identifier (URI) reference.

We give in Figure 2.3 the XML Schema for our example document[5].

The `<schema>` element is the first element in an XML Schema. The prefix `xsd` is used to indicate the actual XML Schema namespace[6], specifying `element`, `attribute`, `string`, `complexType`, etc. The corresponding URI is `http://www.w3.org/2001/XMLSchema`.

The element `library` is of complex type (`complexType`), which is defined as `libraryinfo`. The latter is defined as a sequence of elements of type `bookinfo`, itself a complex type, made of a title, an author, a year and an abstract. The latter three do not contain other elements and attributes, and as such are of basic type, which include string, date, integer, Boolean, decimal, etc. In our example, they are all defined as strings (`string`). Finally, the sub-element title has a complex type because it has an attribute (`isbn`), with default value set to 0.

[4]Here type is with respect to its sense in programming language (e.g. string, integer, array, etc), and not as the type of an element as given by its tag name.

[5]A complete description of XML Schema can be found in the various W3C Recommendation documents available from http://www.w3.org/XML/Schema, under Specifications and Development.

[6]http://www.w3.org/TR/REC-xml-names/

```
<xsd:schema xmlns:xsd="http://www.w3.org/2001/XMLSchema">
   <xsd:element name="library" type="libraryinfo"/>
   <xsd:complexType name="libraryinfo">
      <xsd:sequence>
            <xsd:element name="book" type="bookinfo"
                 maxOccurs="unbounded"/>
      </xsd:sequence>
  </xsd:complexType>
  <xsd:complexType name="bookinfo">
      <xsd:sequence>
            <xsd:element name="title" type="titleinfo"/>
            <xsd:element name="author" type="xsd:string"/>
            <xsd:element name="year" type="xsd:string"/>
            <xsd:element name="abstract" type="xsd:string"
              minOccurs="0"/>
      </xsd:sequence>
  </xsd:complexType>
  <xsd:complexType name="titleinfo" mixed="true">
   <xsd:attribute name="isbn" type="xsd:string" default="0"/>
  </xsd:complexType>
</xsd:schema>
```

Figure 2.3: XML Schema for our example XML document

Two occurrence indicators are used to define how often an element can occur within its parent element. The `<maxOccurs>` indicator specifies the maximum number of times an element can occur, whereas the `<minOccurs>` indicator specifies the minimum number of times an element can occur. Their default values are for both 1. In our example, `maxOccurs="unbounded"` means that the `library` element has an unlimited number of `book` elements as children. Furthermore, since neither `<maxOccurs>` nor `<minOccurs>` have been specified for the `title`, `author` and `year` elements, these appear exactly once in a `book` element. Finally, the `abstract` element can appear, at most, once within a `book` element as the value of `minOccurs` is set to 0.

As for DTDs, an XML Schema provides information, in this case more comprehensive, that may inform the development of an XML retrieval system.

2.5 XML DOCUMENTS AS TREES

Let us consider the XML document shown in Figure 2.4. This is an example of a document-oriented XML document.The logical structure of this document can be described as follows: the

```
<article>
<sec>
  <subsec>
    <p> ... wine ... patagonia ... </p>
    <p> ... wine ... </p>
    <p> ... patagonia ...  </p>
  </subsec>
  <subsec>
    <p> ... </p>
    <p> ... </p>
  </subsec>
</sec>
<sec>
  <p> ... </p>
  <p> ... wine ...  </p>
  <p> ... </p>
  <p> ... </p>
</sec>
</article>
```

Figure 2.4: Sample XML document

XML document corresponds to an article made of two sections; the first section is composed of two subsections; each of these are composed of paragraphs, three in the first subsection and two in the second subsection; finally, the second section is composed of four paragraphs.

This logical structure can be, and is conventionally, viewed as a tree. The tree corresponding to our sample XML document is shown in Figure 2.5, where elements are given unique names so we can refer to them.

This XML document is made of fourteen elements. The `<article>` element is the root of the tree, and the paragraph elements form the leaves of the trees. The tree has a maximum of four levels, thus a depth of 4. The depth of the section element `sec1` is 3, whereas that of `sec2` is 2.

The goal of an XML retrieval system is to identify the most relevant elements to a given query, and to return them as answers to the query. These relevant elements could include the root of the tree, here `article`, if everything in the document is relevant, `sec1` if everything in that section is relevant and nothing in `sec2` is, or one or several paragraphs. For example, returning `para112`, `para113` and `para21` would mean that the XML retrieval system has found relevant content in these paragraphs but none in the other paragraphs.

In other words, the goal of an XML retrieval system is not only to find relevant elements, but those at the right level of granularity, or when viewed as a tree, those at the right level of the tree.

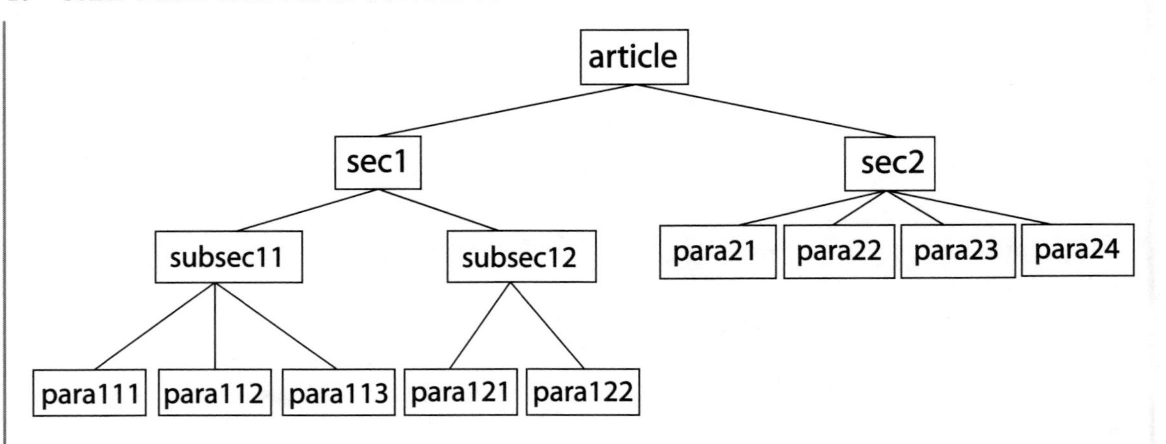

Figure 2.5: The tree view of our sample XML document. Note that the names of the elements, e.g. `subsec12`, are here for display so that we can refer to the elements; they are not part of the XML mark-up

In this book, we describe approaches developed to achieve this goal. Before this, we provide some historical perspectives on areas of research that have influenced XML retrieval.

CHAPTER 3

Historical perspectives

Information retrieval is concerned with the representation, organization, analysis, storage, access and presentation of information items. Although advances in modern multimedia technologies mean that information items not only include text, but also content in other media (e.g., image, speech, audio, video), information available on e.g. the web, in digital libraries, and in intranets remains prevalently textual.

Textual information can be broadly classified into two categories, unstructured and structured. Unstructured information mostly consists of a sequence of words, put together to form sentences, themselves forming paragraphs, with the intention to convey some meanings. Classical information retrieval has been mainly concerned with unstructured information.

Text documents, however, often display structural information. Two types of structure have often been distinguished: layout and logical. The layout structure describes how a document looks like (e.g. a book made of pages and a page formatted into columns) or should be displayed (e.g. section title in bold with a specific font size), whereas the logical structure describes document parts (e.g. an abstract, a paragraph), as well as the relationships between documents parts (e.g. a scientific article made of an abstract, several sections and subsections, each of which composed of paragraphs).

Both types of structural information are usually expressed through a mark-up language. HTML and XML are examples of mark-up languages used to specify the layout and the logical structure, respectively[1]. Unlike HTML, XML can be used to specify semantic information about the stored content. Indeed, a document marked-up in XML has a structure that explicitly identifies semantically separate document parts. This can be exploited to provide powerful access to textual information.

Exploiting the document structure, whether this structure is implicit (e.g. a sequence of paragraphs) or explicit (e.g. as marked up with a language such as XML) means that an information item may correspond to an arbitrary granularity component of a document, e.g. the whole document, a section, several contiguous paragraphs, a sentence. Exploiting the document structure thus allows to break away from the traditional notion of a fixed unit of retrieval and forms the basis of "structured document retrieval".

3.1 STRUCTURED DOCUMENT RETRIEVAL

Structured document retrieval is concerned with the development of the retrieval approaches, where document components, instead of whole documents, are returned to the user in response to a query.

[1] It should be noted that the successor of HTML, XHTML, has introduced significant changes in terms of flexibility and is a step towards separating the logical structure (sub-division in logical units) from the layout structure. See http://www.w3.org/TR/xhtml1/.

This retrieval paradigm is of particular benefit for collections containing long documents or documents covering a wide variety of topics (e.g., books, user manuals, legal documents, etc.), where the users' effort to locate relevant content can be reduced by directing them to the relevant parts of documents.

The term "structured document retrieval" was introduced in the early to mid 90s in the information retrieval community. At that time, it referred mainly to "passage retrieval" (Section 3.2) and "structured text retrieval" (Section 3.3).

In the late 1990s, the interest in structured document retrieval grew significantly due to the introduction of XML in 1998. As XML has become the standard for document mark-up, an increasing number of documents are being made available in this format. As a consequence numerous techniques have been and are being developed to represent, organize, analyze, store, access and present XML documents. Research on "XML retrieval" was then further boosted with the set-up of INEX in 2002, the Initiative for the Evaluation of XML Retrieval (Section 3.8), which allowed researchers to compare and discuss the effectiveness of models specifically developed for XML retrieval [65]. Nowadays, XML retrieval is almost a synonym for structured document retrieval.

3.2 PASSAGE RETRIEVAL

Passage retrieval has been studied in information retrieval around the mid 90s, e.g., [24, 70, 87, 158, 201, 131]. In passage retrieval, documents are first decomposed into passages, where a main research issue is the decomposition of the document into "meaningful" passages. Three main techniques were proposed, namely, fixed-size text-windows of words, fixed discourses such as paragraphs, or topic segments through the application of a topic segmentation algorithm. Passages could themselves be retrieved as results to a query, or be used to rank documents as results to the query. Since 2007, INEX has a passage retrieval task, where the aim is to identify the appropriate size of results to return and their location within the document. This is to be contrasted to the previous INEX tasks, where elements, i.e., components explicitly delineated by XML tags, are retrieved.

It should be noted that passage retrieval can be used in other processing tasks, where there is the need to first identify relevant passages in documents, for example in question answering systems.

3.3 STRUCTURED TEXT RETRIEVAL

Structured text retrieval is concerned with the developments of models for querying and retrieving from structured text, where the structure is usually encoded with the use of mark-up languages, such as SGML, and now predominantly XML. Structured text retrieval has considered both types of structure, layout and logical, although with a stronger emphasize on the latter. The use of the term "structured" in "structured text retrieval" is there to emphasize the interest in the structure.

In the end of the 80s and throughout the 90s, various structured text retrieval models have appeared in the literature, e.g. [63, 23, 133, 31, 120, 102, 33]. These can be viewed as precursor of approaches specifically developed for XML retrieval. They can be contrasted by their structuring power [71], namely, explicit vs. implicit, static vs. dynamic, and single vs. multiple hierarchical.

Structured text retrieval models work mostly on the basis that the documents are structured, through the use of a mark-up language, where it is clear which portions of the documents correspond to a section, a paragraph, a title, and so on. Few assume an implicit structure, where documents are modeled as sequences of tokens without distinguishing a word token from a mark-up token. A structural element is, therefore, constructed at querying time by looking up the opening and closing mark-up tokens. Those regions starting with proper opening tokens, ending with corresponding closing tokens, and satisfying the content conditions are returned.

A text retrieval model that allows the specification of dynamic structures in the query is able to return elements or regions that have not been explicitly marked-up in the text of the documents. This is the case with current XML query languages, e.g. XQuery (described in Section 4.3), which allow for the construction of new elements.

Finally, the type of structure most used with structured text retrieval models is hierarchical. There are, however, approaches that can deal with multiple structural hierarchies on the same document e.g. [23, 63, 133, 3], where each hierarchy provides a different view of the document. For instance, one hierarchy might represent the logical structure of the text (chapters, sections, subsections), whereas a second hierarchy might represent the layout structure (columns, pages).

Overall, structured text retrieval models are more flexible and expressive than models developed specifically for XML retrieval. However, many, if not most, of them do not return a ranked list of results, as this was not their main concerns. Their primary aims were mainly the expressiveness of the query language and efficiency.

3.4 HYPERTEXT AND WEB RETRIEVAL

Structured document retrieval should be contrasted to hypertext, e.g. [1], and web retrieval. There, documents are linked according to what can be referred to as an external structure, as opposed to the internal structure of a document, where the components of the document are linked. A major difference between internal and external structures is that the former is most of the time restricted to a tree structure, while the latter may form an arbitrary graph (e.g., a web graph). Structured document retrieval, and now XML retrieval, have mainly been concerned with structured documents and their associated inner logical structure. There is, however, work that looks at exploiting links across documents as an additional source of evidence for improving XML retrieval retrieval performance, e.g. [103].

3.5 DATA- VS DOCUMENT-CENTRIC XML DOCUMENTS

There are two different views on XML, namely, data-centric and document-centric.

Data-centric documents are characterized by a fairly regular structure, fine-grained data (that is, the smallest independent unit of data is at the level of a PCDATA-only element or an attribute),

```
<?xml version="1.0" encoding="UTF-8" standalone="no"?>
<!DOCTYPE CLASS SYSTEM "class.dtd">
<CLASS name= "DCS317" num_of_std="100">
     <LECTURER lecid="111">Mickey Mouse</LECTURER>
     <STUDENT marks="70" origin="Oversea">
            <NAME>James Bond</NAME>
    </STUDENT>
    <STUDENT marks="30" origin="EU">
            <NAME>Donald Duck</NAME>
    </STUDENT>
</CLASS>
```

Figure 3.1: Example of a data-centric XML document

and little or no mixed content. The order in which sibling elements and PCDATA occurs is generally not important, except when validating the document. Data-centric documents originate both in a database and thus needs to be presented in XML, or outside a database and thus needs to be stored in a database. An example of the former is the vast amount of legacy data stored in relational databases, whereas an example of the latter is scientific data gathered by a measurement system and converted to XML. An example of a data-centric XML document is given in Figure 3.1.

Data-centric documents are documents designed for machine consumption. Data-centric approaches to XML retrieval are concerned with the efficient representation and querying of XML documents from a database perspective, and are referred to as "querying semi-structured data" (Section 3.6).

Document-centric documents are (usually) documents that are designed for human consumption. They are characterized by a less regular or irregular structure, larger grained data (that is, the smallest independent unit of data might be at the level of an element with mixed content), and large amount of mixed content. The order in which sibling elements and PCDATA occurs is usually important. Unlike data-centric documents, document-centric documents usually do not originate from a database. The XML markup serves mainly as a means for exposing the logical structure of document texts. Figure 3.2 shows an example of a document-centric document.

Approaches to retrieve document-centric documents are built on concepts developed in information retrieval and extend these to deal with structured information. This book is concerned with XML retrieval, as being investigated by the information retrieval community, i.e. how to effectively access document-centric corpora, referred to as "content-oriented XML retrieval" (Section 3.7).

It should be pointed out that research in information retrieval and databases on XML retrieval have been concerned, because of historical reasons, with different aspects of the retrieval process,

```
<?xml version="1.0" encoding="UTF-8" standalone="yes"?>
<CLASS name="DCS317" num_of_std="100">
<LECTURER lecid="111">Mickey Mouse</LECTURER>
<STUDENT studid="007" >
<NAME>James Bond</NAME> is the best student in the
class. He scored <INTERM>95</INTERM> points out of
<MAX>100</MAX>. His presentation of <ARTICLE>Using
Materialized Views in Data Warehouse</ARTICLE> was
brilliant.
</STUDENT>
<STUDENT stuid="131">
<NAME>Donald Duck</NAME> is not a very good
student. He scored <INTERM>20</INTERM> points ...
</STUDENT>
</CLASS>
```

Figure 3.2: Example of a document-centric XML document

e.g. ranking in information retrieval versus efficiency in databases. Nowadays, there is a convergence trend between the two areas (e.g. [6, 8, 14, 34]).

3.6 QUERYING SEMI-STRUCTURED DATA

The term "semi-structured" comes from the database community. Traditional database technologies, such as relational databases, have been concerned with the querying and retrieval of highly structured data (e.g. from a student table, find the names and addresses of those with a grade over 80 in a particular subject). Text documents marked-up, for instance, in XML are made of a mixture of highly structured components (e.g. year, author name) typical of database records, and loosely structured components (e.g. abstract, section). Database technologies are being extended to query and retrieve such loosely structured components, called semi-structured data. Databases that support this kind of data, mainly in the form of text with mark-up, are referred to as semi-structured databases, to emphasize the loose structure of the data and use "querying data" instead of "data retrieval". The main research issues are the efficient representation and querying of data-centric XML corpora, e.g., [20].

3.7 CONTENT-ORIENTED XML RETRIEVAL

Content-oriented XML retrieval focuses on the document-centric view of XML. Following the goals of structured document retrieval, XML retrieval aims to exploit the available structural information in documents in order to implement a retrieval strategy that returns to the user document components of a particular type, e.g. an abstract – instead of complete documents. For example, in response to a user query on a collection of digital textbooks marked-up in XML, an XML retrieval system may return a mixture of paragraph, section and chapter elements that have been estimated to best answer a query.

Content-oriented XML retrieval can hence be viewed as a special case of structured document retrieval in the sense that XML retrieval focuses on documents marked-up in XML, while structured document retrieval builds on a more general notion of structural information, e.g. [132, 18, 27]. Although structured document retrieval predates the development of XML, it is the widespread use of XML as a standard document format that has made structured document retrieval a hot topic of research. In the remaining of this book, we simply use XML retrieval to refer to content-oriented XML retrieval.

3.8 INITIATIVE FOR THE EVALUATION OF XML RETRIEVAL (INEX)

Recent advances in XML retrieval were made possible by the set-up of the INitiative for the Evaluation of XML retrieval. INEX was set up in 2002 to address the need for a common evaluation benchmark for assessing the retrieval performance of XML retrieval systems. INEX has been promoting research in XML retrieval by providing a forum for researchers to evaluate their XML retrieval approaches and compare their results [47, 52, 53, 54, 55, 56]. Chapter 8 contains a detailed description of the test collections and evaluation methodologies developed at INEX.

3.9 FOCUSED RETRIEVAL

Structured document retrieval, passage retrieval, structured text retrieval, querying semi-structured data, XML retrieval, all belong to what has recently been named "focused retrieval" [182, 183][2]. The latter is concerned with returning the most focused results to a given query. In XML retrieval, these focused results consist of XML elements, those that are relevant to a user information need, and that are specific to that information need. This book describes in details how such focused results are computed.

[2]Focused results also include factoid answers as would be the case with question & answering systems (e.g., "What is the capital of the UK" e.g. [195]).

CHAPTER 4

Query languages

The appearance of XML sparked numerous debates regarding the functionalities of a query language for XML retrieval, see for example the views and developments presented at the Query Language workshop in 1998 (QL 98)[1]. These views and developments were initially mostly concerned with querying data-centric XML documents. Examples of resulting XML query languages include XML-QL [43], XSLT[2], XQL[3], which led to XPath[4], Quilt [26], and XQuery[5]. The latter is the W3C adopted query language for (data-centric) XML retrieval, and was influenced by many of the other mentioned query languages.

Later developments focused on extensions or adaptations of data-centric query languages for content-oriented XML retrieval, in order to account for word search that goes beyond string matching and ranking. Examples of resulting query languages are NEXI [186] based on XPath and developed by the INEX community, ELIXIR [28], which extents XML-QL, XIRQL [49], which extends XQL, and XQuery Full-Text [5] [6], which extends XQuery.

We do not provide a detailed overview or a comparison of query languages for XML retrieval (an excellent survey on XML query languages for data- and content-oriented XML retrieval can be found in [39]). Instead, we discuss the functionalities of a query language for content-oriented XML retrieval. Our particular interests are the added structural constraints to an information retrieval-like query, i.e. of the form "find me documents on a given topic". For this purpose, we first list the main types of structural constraints for XML retrieval (Section 4.1). We then present a classification of XML query languages (Section 4.2), together with the types of structural constraints they encompass. Finally, we describe four XML query languages, namely, XPath, NEXI, XQuery, and XQuery Full-Text (Section 4.3). These query languages provide a good overview of the important developments on XML query languages for document-centric XML retrieval.

4.1 STRUCTURAL CONSTRAINTS

In information retrieval, a user submits a query to specify conditions on the content of a document or a document fragment, e.g. "I want a document (or a document component) on xml retrieval evaluation". We refer to these as content conditions. Various types of content constraints exist. The most common one, which we refer to as word constraint, consists of a list of words specifying that

[1] http://www.w3.org/TandS/QL/QL98/
[2] http://www.w3.org/TR/xslt
[3] http://www.ibiblio.org/xql/xql-proposal.html
[4] http://www.w3.org/TR/xpath
[5] http://www.w3.org/TR/xquery/
[6] http://www.w3.org/TR/xpath-full-text-10/

the document components to be returned should contain or be about these words. Word constraints are the classical input to most information retrieval systems.

Conditions can be imposed on the positions of the words in the text (of document or document components), such as forming a phrase or being within a specified distance (e.g., "information and retrieval with distance 4"). We refer to this type of condition as a context constraint. Also a weight can be used to specify the importance of words and/or context constraints in the document components, which forms what we call a weight constraint. For instance, "0.6 xml 0.2 retrieval" means that "xml" is more important that "retrieval" when deciding which document components to return as results, and "+xml retrieval" means that a document component must definitively contain or be about "xml" to be returned. Finally all the above can be combined using Boolean operators, forming a Boolean condition. For instance, "0.6 xml or (information retrieval with distance 4)". The document components that satisfy the Boolean expression are returned as results.

In (traditional) databases, processing the content constraints of a query yields (mostly) a non-ranked list of document fragments, whereas in information retrieval, the list is ranked (so that to present first the best matching components to the users). Here, document fragments refer to either document components or XML elements as they appear in a document, or new XML elements that are constructed to form a result. We return to this later. Also, in databases, it is commonly the case that the words must be contained in a document component, whereas in information retrieval, containment is replaced by aboutness, i.e., a document component must be about (the topic described by) a word.

XML allows the specification of structural constraints in addition to content constraints. Indeed, when searching XML documents, users have the possibility to think of more precise queries, for example, "I want a section discussing xml retrieval evaluation contained in a chapter on evaluation initiatives". Here, "xml retrieval evaluation" and "evaluation initiatives" specify constraints with respect to the content to be retrieved (as discussed above), and "section", "chapter", and "section in chapter" specify structural constraints on the units to retrieve.

This above example illustrates the idea of a query language that would allow combining the specification of the desired textual content with the specification of structural constraints. For content-oriented XML retrieval, there are three main types of structural constraints [9].

The first one allows the specification of the target result. If the structure of the desired results is known, then users have the opportunity to specify which particular structural results they are targeting, e.g., the user requests "abstract" components from a collection of scientific articles.

Structure can be used to specify structural constraints other than that of the desired results. For instance, a query requesting "sections about xml retrieval evaluation from documents with abstracts about evaluation initiatives" contains two structural constraints: one with respect to the sections (about "xml retrieval"), specifying the target results as just mentioned, and the second with respect to other components of the document (the abstract being about "evaluation initiatives"), providing additional structural constraints. The latter is referred to as a support condition.

Users may want to obtain as answers to their queries not only existing components (as they appear in the document), but answers that are built from several components within or across documents. For example, a user may wish to obtain "the title of a section, with its first and last paragraph grouped together into one fragment". This type of structural requests, which is common in database, is referred to as result construction.

XML query languages have mainly been concerned with the specification of some or all of the types of content and structural constraints as above described. It should, however, be noted that an increase in expressiveness entails an increase in complexity, both in terms of translating user information needs into correct queries, so that they can be processed, and of the actual processing of the queries.

4.2 CLASSIFICATION OF XML QUERY LANGUAGES

We recall that in XML retrieval, the document structure can be used to determine which document components or fragments are more meaningful to return as answers to a query. The structure can also be used to specify in addition conditions on the structure with the aim to limit the search to specific XML elements, as opposed to whole documents. XML query languages, thus, can be classified as content-only or content-and-structure query languages.

4.2.1 CONTENT-ONLY

Content-only queries make use of content constraints to express user information needs. In their simplest form, they are made of words, which historically have been used as the standard form of input in information retrieval. They are suitable for XML retrieval scenarios in which users do not know or are not concerned with the document structure when expressing their information needs. Although only the content aspect of the information need is being specified, XML retrieval systems must still determine what are the best fragments, i.e., the XML elements at the most appropriate level of granularity, to return as answers to a query.

4.2.2 CONTENT-AND-STRUCTURE

Content-and-structure queries provide a means for users to specify their content and structural information needs. It is towards the development of this type of queries that most research on XML query languages lies. These languages possess some of the content and structure characteristics listed in Section 4.1. Here, we discuss mainly how the structure characteristics are captured by these languages. Section 4.3 contains a detailed description of several XML query languages for content-and-structure queries.

There are three main categories of content-and-structure query languages, namely tag-based languages, path-based languages, and clause-based languages, where the complexity and the expressiveness of these query languages increase from tag-based to clause-based queries. From a user perspective, this increase in expressiveness and complexity often means that content-and-structure queries are hard to write. Nonetheless, they can be very useful for expert users in specialized scenarios, such as patent retrieval and genomic search.

4.2.2.1 Tag-Based Queries

These queries allow users to annotate words in the query with a single tag name that specifies a structural constraint to target as the result. For example, the information need "retrieve sections about xml retrieval" would be expressed as follows:

```
section: xml retrieval
```

An example of a tag-based query language is XSEarch [32].

Tag-based queries are intuitive, and have been used in domains outside XML retrieval (e.g. faceted search, web search). However, they only express simple, although important and likely to be common, structural constraints. They lack the expressiveness of path-based and clause-based query languages, as they do not cater for support conditions and result constructions.

4.2.2.2 Path-Based Queries

These types of queries are based upon the syntax of XPath to encapsulate the document structure in the query. Examples of path-based query languages include XPath[7], XIRQL [50], and NEXI [186]. For example, the information need "retrieve sections about xml retrieval in documents about information retrieval" would be expressed as follows in NEXI:

```
//document[about(., information retrieval)]
                 //section[about(., xml retrieval)]
```

Path-based queries allow for expressing target results ("section" element above) and support conditions (the "section" should be contained in a "document " about "information retrieval"), but they do not allow the construction of new fragments.

It should be noted that any tag-based query, for example, `section: xml retrieval`, can be re-written using a path-based query language, for example, in NEXI as follows:

```
//section[., about(xml retrieval)]
```

Moreover, any content-only query can be expressed as a path-based query, for example as follows in the NEXI query language:

```
//*[., about(xml retrieval)]
```

which asks for any element, at any level of granularity, on the topic of "xml retrieval".

4.2.2.3 Clause-Based Queries

These queries use nested clauses to express information needs, very similarly to SQL (Structured Query Language) in databases. The most prominent clause-based language for XML retrieval is

[7] http://www.w3.org/TR/xpath

XQuery[8]. A typical clause-based query is made of three clauses, "for", "let" and "return", each with a role; for instance, in our context: the "for" clause to specify support conditions; the "where" clause to specify content constraints; and the "return" clause to specify target elements and the construction of new fragments as result. Within the "for" and "return" clauses, XPath expressions are used to refer to the structure of the document. The following information need "retrieve document sections with the title xml retrieval" would be expressed as follows in XQuery:

```
for $x in /document/section
   where $x/title="xml retrieval"
   return $x
```

XQuery Full-Text [5] extends XQuery with powerful text search operations, including context constraints described in Section 4.1 (e.g., proximity distance) and ranking functionality.

4.3 EXAMPLES OF XML QUERY LANGUAGES

We describe now four content-and-structure query languages. As discussed in the previous section, XML content-only query languages are specified in the same way as in flat document retrieval, and as such are not further discussed. We present here two path-based query languages, namely XPath and NEXI, and two clause-based query languages, namely XQuery and XQuery Full-Text.

4.3.1 XPATH

XPath (XML Path language) is a query language defined by the W3C. Its primary purpose is to access or navigate to components of an XML document. In addition, XPath provides basic facilities for the manipulation of strings, numbers and Booleans. The first working draft for XPath (XPath 1.0) was published in July 1999, which resulted in its recommendation[9] in November 1999[10].

The most important type of expressions in XPath is the location path, which consists of a series of navigation steps within an XML document. `book/publisher/@isbn` is a location path, where `book` and `publisher` are steps that navigate to (or select) children elements with names "book" and "publisher", respectively, and `@isbn` is a step that navigates to attributes with name "isbn". All the steps are separated by "/", which means that the location path selects the `isbn` attributes that are directly below `publisher` elements, themselves directly below `book` elements. `book/publisher` corresponds to what is called a child axis step in XPath.

The navigation steps can be separated by "//", which means that the location path navigates to the current element and all its descendant elements before it applies the next step. For example, `book//title` navigates to all title elements that are directly or indirectly below a book element, whereas `//title` will select all title elements in the document.

[8] http://www.w3.org/TR/xquery/

[9] http://www.w3.org/TR/xpath

[10] There is a current recommendation for XPath 2.0 (http://www.w3.org/TR/xpath20/). Here we only describe XPath 1.0, as we are mostly interested in the details of the location path, but see footnote 13.

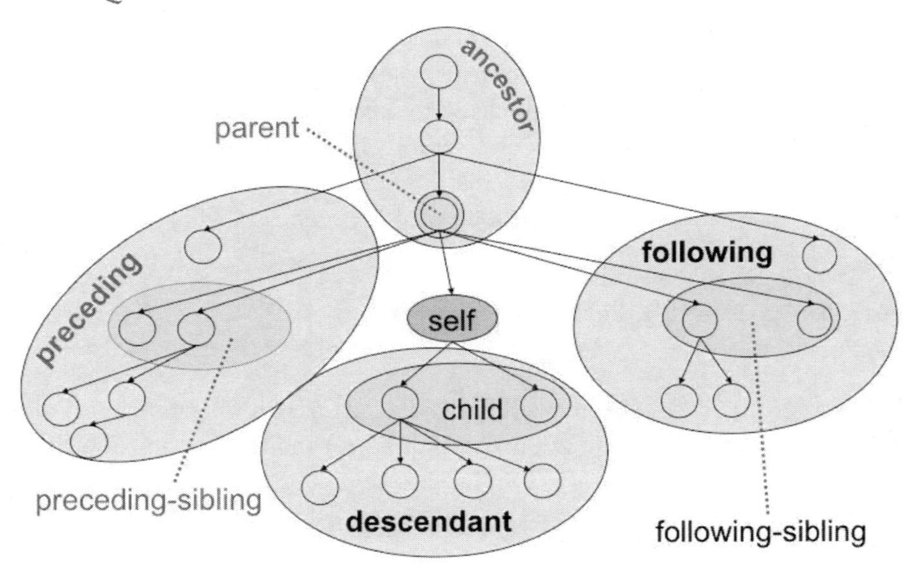

Figure 4.1: Examples of axes in XPath (Courtesy of Benjamin Piworwarski)

Special steps include the self step denoted "." and parent step "..". For example, `.//book` returns any book elements contained in the current element, whereas `../publisher` returns the publisher elements of the parent of the current element. Also, XPath uses wildcards such as "*" and "@*" to navigate to elements and attributes with any name, e.g. `book/*` and `book/publisher/@*`.

At each step, predicates can be specified between "[" and "]", which must be satisfied for elements to be selected (navigated). For example, the following XPath expression `//book[@year=2002]/title` selects the title of books published in 2002, and only those. The standard comparison operators =, ! = , < and <= can also be used in the predicates.

Existential predicates are used to check whether a certain path expression returns a non-empty result. For example, the following XPath expression `//publisher[city]` selects publishers for which the city information is given.

Finally, positional predicates are used to navigate according to the position of an element in the document tree. For example, `//publisher/country[1]/city` selects for each publisher the cities of the first listed country (we assume that a publisher is based in several countries). The comparisons and existential conditions can be combined with "and" and "or" operations, and the "not()" function, e.g., `not(@year = 2002)`.

So far, the navigation steps have been with respect to child/parent or more generally ancestor/descendent movements. For more complex navigation steps a formal syntax is used, in which a single step is specified in the form of `axis::node-test` followed by zero or more predicates. For instance, the `following` axis is used to navigate to all elements in the same document that are after the current element and following the order of the elements in the document. The `following-sibling` axis is used to navigate to all the following siblings of the current element. We illustrate in Figure 4.1 axes defined in XPath with respect to the element referred to as `self`. We can see that this element has five elements on the `following` axis, two of which are on the `following-sibling` axis. These numbers for the `preceding` axis and `preceding-sibling` axis are six and two, respectively. Similarly, the element has three and six elements on the `ancestor` axis and the `descendant` axis, respectively, meaning that the element has three ancestors and six descendants. Finally, it should be noted that, for example, the location path `section/paragraph` is actually an abbreviation of the formal expression `child::section/child::paragraph`, as all location paths are formally defined using the `axis::node-test` syntax. In this abbreviation, `child::` is omitted.

An important function in XPath for the purpose of content-oriented XML retrieval is the Boolean function `contains()`, which takes two string arguments and returns true if the first string contains the second string, and false otherwise. This function can be used to check whether an element contains in its text (as specified by the first argument) a specified string (as specified by the second argument). The outcome is, however, not a ranked list of elements, but a set of elements. As such, XPath is not a query language that can be directly used for content-oriented XML retrieval. XPath is, however, used by other XML query languages or has inspired other XML query languages, some of which allowing for the ranking of results. We discuss one such language next.

4.3.2 NEXI

The Narrowed Extended XPath I (NEXI) query language [186] was developed at INEX, as a simple query language for content-oriented XML retrieval evaluation. NEXI, which consists of a small but enhanced subset of XPath, has been used by INEX participants to express realistic content-and-structure queries used to form the INEX test collections. NEXI was based on XPath because the latter is a language already well understood by the XML community, and it would have been a disadvantage for INEX to propose another XML language.

The enhancement comes from the introduction of a new function, named `about()`. The XPath `contains()` function, which requires an element (its text) to contain the given string content, was replaced by the `about()` function, which requires an element to be about the content. This is to reflect that an element can be relevant to a given query (its content aspect) without actually containing any of the words used in the query.

The reason for choosing a small subset of XPath were two-fold. First, the analysis of the submitted queries to INEX showed high syntactic and semantic error rates in the use of XPath path location to express structural constraints [136]. As a result, all location paths that were deemed unnecessary for evaluating content-oriented XML retrieval effectiveness were removed. For instance, the parent/child navigation step was considered particularly problematic as it was open to misinterpretation, and hence was removed.

The second reason was that NEXI was developed for the purpose of evaluating the effectiveness of XML retrieval systems. For this reason, positional predicates (e.g. `//paragraph[1]`) are not allowed, as they do not bring anything in terms of comparing the effectiveness of XML retrieval systems. Also, all target elements must have at least one content condition, i.e., one `about()` function. It is indeed a mechanical process to return, for instance, the title of sections on a given topic. For the purpose of evaluating retrieval effectiveness, what matters is that the relevant sections for the given topic are actually returned.

An example of a NEXI query is:

```
//article[about(.//bdy, "information retrieval")]//
                    section[about(., xml) and about(., retrieval)]
```

The target result is `//article//section`. Several content constraints are used, e.g., `about(.//bdy, "information retrieval")`, including one for the target result. A Boolean operator is also used to separate `about(., xml)` and `about(., retrieval)`. NEXI also allows for weighted constraints ("+" to emphasize the importance of word and "-" to emphasize the opposite effect), and one type of context constraints, i.e. phrases such as `"information retrieval"` in double quotes above.

NEXI was developed by the INEX community to construct topics for the purpose of evaluating XML retrieval effectiveness. NEXI offers a good compromise between the need to formally express content and structural constraints, and the ability to write information retrieval-like queries.

It still remains the task of the XML retrieval system to interpret a NEXI query. The first interpretation is with respect to the `about()` condition as implemented by the retrieval model, and the second is with respect to the structural constraint, as implemented by the query processing engine, used by the XML retrieval system. Chapter 6 describes approaches used to implement the about conditions, and, in particular, Section 6.6 describes approaches used to process structural constraints, for the purpose of ranking XML elements for given queries.

4.3.3 XQUERY

XQuery is an XML query language that includes XPath as a sub-language, but adds the possibility to query multiple documents and combine the results into new XML fragments (result construction). XQuery is the result of a W3C working group established as the result of a workshop on query languages for XML[11]. The Recommendation document for XQuery 1.0[12] was published in January 2007[13].

Most features of XQuery can be traced back to its immediate predecessor Quilt [26], which was created to query heterogeneous data sources, and which also draws on the design of several languages, for instance XPath, from which the concept of path expressions was taken to navigate in hierarchically structured documents. From XML-QL (semi-structured information oriented language) [43] came the idea of using variable bindings to construct new fragments to be returned as results. Older

[11] http://www.w3.org/TandS/QL/QL98
[12] http://www.w3.org/TR/xquery/
[13] In January 2007 the recommendations for both XQuery 1.0 and XPath 2.0 were published together. XPath 2.0 is based on the same data model as XQuery 1.0 and is semantically and syntactically a subset of XQuery 1.0.

influences include SQL with its "select-from-where" clauses to restructure data and operations such as join and groupings, which formed the inspiration for the FLWOR expressions, the core expression of XQuery.

To illustrate FLWOR, consider the following example of an XQuery expression that requests a list of publishers, ordered by name, whose average price of books is less than 50 pounds:

```
for $pub in distinct-values (doc("pub.xml")//publisher)
 let $a := avg(doc("bib.xml")//book[publisher = $pub]/price)
      where $a < 50
      order by $pub/name
      return
          <publisher> { $pub/name , $a } </publisher>
```

A FLWOR expression starts with one or more `for` and `let` clauses, each binding a number of variables (starting with $). The `for` clause is used to create a sequence of elements, where the bound variable is used to iterate over the elements of that sequence. The `let` clause is used to bind a sequence to a variable. An optional `where` clause specifies conditions on the values of the variables, and an optional `order by` clause provides a sorting criteria. Finally, a `return` clause specifies the results to be returned.

In the above example, the `for` clause binds the variable $pub such that it iterates over the publisher elements in the document entitled "pub.xml" in the order that they appear. The `distinct-values` function eliminates duplicates in "pub.xml". For every binding of the variable $pub, the `let` clause binds the variable $a to the average price of books from publisher $pub. Then, those publisher elements for which the condition in the `where` clause is true are selected, i.e., those for which the average price as assigned to $a is less than 50. The resulting bindings are sorted by the `order by` clause on the publisher name in $pub ($pub/name). Finally, the `return` clause creates, for each binding $pub and $a in the result of the preceding clause, a new publisher element that contains the name element of the publisher $pub, and the associated average price $a. Without the `order by` clause, the results would have been sorted in the order in which the publisher elements appear in "pub.xml". The results are new fragments, as they were not in the XML original documents, and are generated using the sequence operator ",".

XQuery is a powerful query language for XML retrieval, and can be viewed as the SQL for XML (it is indeed a superset of SQL). It is a language that is mostly appropriate for data-centric XML retrieval. This is because its text search capabilities are limited and, in addition, it does not provide any ranking of results, the latter being crucial in content-oriented XML retrieval. These shortcomings led to the specification and development of the query language discussed next.

4.3.4 XQUERY FULL-TEXT

XQuery Full-Text [5] is an XML query language that extends XQuery with powerful text search capabilities. For example, with XQuery Full-Text, a user can ask for target elements that "contain the words xml and retrieval within three words of each other, ignoring stemming variations of the word evaluation", which is not possible to express with XQuery. Furthermore, and more importantly with respect to content-oriented XML retrieval, XQuery Full-Text allows to specify that the results should be ranked according to how relevant they are.

XQuery Full-Text has been designed to meet the "XQuery 1.0" and "XPath 2.0 Full-Text 1.0" Requirement [14] and Use Cases [15], and has been inspired by earlier query languages for document-centric XML retrieval such as ELIXIR [28], JuruXML [25], and XIRQL [50]. The document upon which this section is based was published in May 2008[16].

The added text search capabilities are the result of the introduction of a new XQuery expression, *FTContainsExpr*, which acts as a regular XQuery expression fully composable with the rest of XQuery and XPath expressions. For instance, the following *FTContainsExpr* expression:

```
//book[./title ftcontains {"xml",  "retrieval"} all]//author
```

returns the authors of books whose title contains all the specified words, here "xml" and "retrieval".

XQuery Full-Text defines primitives for searching text, such as phrase, word order, word proximity (i.e., content context constraints). It also allows the specification of letter cases in matched words, the use of stemming, thesauri, stop words, regular-expression wildcards, and many more. For instance, the following *FTContainsExpr* expression restricts the proximity of the matched words to appear within a window of six words:

```
//book[./title ftcontains {"xml",  "retrieval"}
        all window 6 words]//author
```

The *FTContainsExpr* expression below looks for matches to the word "evaluation" in its various forms, e.g. "evaluate", "evaluates":

```
//book[./title ftcontains "evaluation" with stemming]//author
```

Note that in information retrieval, stemming is a common process applied at indexing time.

The ranking of results is provided with the introduction of *FTScoreClause* expressions, which allow the specification of score variables. These provide access to the scores of the results of the evaluation of an *FTContainsExpr*. We illustrate with an information retrieval-like example:

```
for $b score $s in //book[./title ftcontains
                             {"xml", "retrieval"} all]
      order by $s descending
      return <book isbn= "{$b/@isbn}", score="{$s}"/>
```

The above query iterates over all books whose title contain both "xml" and "retrieval", where the `$b` variable binds the score of each such book to the score variable `$s`. Both variables are used to return the "isbn" numbers of the books and their scores in order of decreasing relevance.

[14]http://www.w3.org/TR/2007/WD-xpath-full-text-10-requirements
[15]http://www.w3.org/TR/2007/WD-xpath-full-text-10-use-cases
[16]http://www.w3.org/TR/xquery-full-text/

XQuery Full-Text was not designed to implement a specific scoring method, but to allow an implementation to proceed as it wishes. In other words, the above query does not specify how the $s values are calculated. Each XQuery Full-Text implementation can use a scoring method of its choice as long as the generated scores are in the range [0, 1], where higher scores denote greater relevance.

4.4 DISCUSSION

A survey in 2002 [39] showed that early query languages for XML documents suffered from serious limitations with regard to content-oriented XML retrieval. Mainly, the ranking of the results by relevance was not possible. As a consequence, several query languages for content-oriented XML retrieval have emerged. Examples include XXL [172], ELIXIR [28], XIRQL [49], and, discussed in this chapter, NEXI and XQuery Full-Text.

XQuery Full-Text possesses all the characteristics required by both data and document-centric XML retrieval applications. It was indeed developed for this exact purpose. It also allows the implementation of scoring functions, which can be used to provide ranking of results. From a content-oriented XML retrieval perspective, XQuery Full-Text may be viewed as far too complex for many users to master, which is one of the reasons the INEX community developed NEXI, a path-based query language with less expressiveness than a clause-based query language, as its query language. A second reason was to keep the construction of the test collections manageable, for instance during the assessment task (see Chapter 8). A third reason is the complexity of XQuery Full-Text for system design; NEXI was more focused on the use of structure rather than on query language complexity.

XQuery Full-Text can, however, be appropriate and in fact required, in applications involving expert users, such as in the medical domain or patent industry. In addition, the ability to construct new results (also possible with XQuery) is of benefit to perform so-called aggregated search [110], i.e., combining several components from the same document or across documents to form new results (although an extra layer of complexity would be needed to deal with novelty and diversity, both crucial to aggregated search).

The current specification of XQuery Full-Text is a W3C Candidate Recommendation. This document is intended to be submitted for consideration as a W3C Proposed Recommendation as soon as a number of conditions are met, for instance that a test suite is available for testing each identified feature, and that the minimal conformance to this specification has been demonstrated by at least two distinct implementations. This document, published on 16 May 2008, is believed to be sufficiently mature and stable to allow for development and implementation[17].

A topic not covered in this book is that of interfaces for XML retrieval (although we touch upon this in Chapter 7, where we describe the presentation of XML retrieval results). It is obvious that appropriate interfaces are needed to effectively support the more complex interaction between users and an XML retrieval system. One particular aspect is how to support users in expressing content-and-structure queries. Some promising work on this aspect is already available. Zwol et al [203] showed that a form-based interface, reflecting the NEXI query language, worked well.

[17] http://www.w3.org/TR/xpath-full-text-10/

Woodley and Geva [202] looked at natural language-based interfaces, where users entered their content-and-structure information needs in plain language, which were then translated into, in their case, NEXI queries. This also worked well. More research is, however, needed to generalize these findings and obtain further insights for example through large-scale and longitudinal user studies.

The next two chapters of this book describe two core tasks of an XML retrieval system, namely, indexing strategies and ranking strategies.

CHAPTER 5

Indexing strategies

To retrieve documents relevant to a query, the first task of an information retrieval system is to index all documents in the collection. The indexing task aims to obtain a representation of the content of each document (i.e. what the document is about), which is then used by the retrieval algorithm to estimate the degree of relevance of the document, through a retrieval score, to a given query. Documents are then ranked on the basis of their retrieval scores, and then returned as a ranked list to users as answers to their queries.

In XML retrieval, in contrast to "flat" document retrieval, there are no pre-defined fixed retrieval units. The whole document, a part of it (e.g., one of its section), or a part of a part (e.g., a paragraph in the section), all constitute potential answers to a given query. It is, therefore, an open question which elements should be indexed, as indexing all elements may be neither efficient nor necessary. As a result, several indexing strategies have been proposed.

Representing the content of documents involves several steps, namely, tokenisation, stop word removal, stemming, and term statistics computation [124]. In XML retrieval, an additional step is the parsing of the XML format. In this chapter, we concentrate on one particular step, the calculation of terms statistics[1]. Here, a term, also called a keyword or indexing term in information retrieval literature, refers to a descriptive item that is extracted from a document text as the outcome of the indexing process. For instance, processing the text fragment "the understanding of XML information retrieval in the context of INEX" would lead to the following terms "understand", "xml", "inform", "retriev", "context" and "inex". A document is indexed by such set of terms, where each term has associated statistics. These term statistics are used by the retrieval algorithm to estimate a document relevance.

The predominantly used term statistics in information retrieval is the within-document term frequency, tf, and the inverse document frequency, idf. Most others are variants or extensions of these. tf is the number of occurrences of a term in a document and reflects how well a term captures the topic of a document. idf is the inverse number of documents in which a term appears and is used to reflect how well a term discriminates between documents. With these term statistics, an index is built, for instance in the form of an inverted file, which gives for each term in the collection its idf, and for each document containing that term, the corresponding tf.

Representing the content of XML documents requires the calculation of the same terms statistics, but at element level. The simplest strategy is to replace document by element and calculate within-element term frequencies, etf, and inverse element frequencies, ief. This, however, raises an issue because of the nested nature of XML documents. For instance, suppose that a section element

[1]The other steps are well explained in many information retrieval books, to name a few [188, 15, 124].

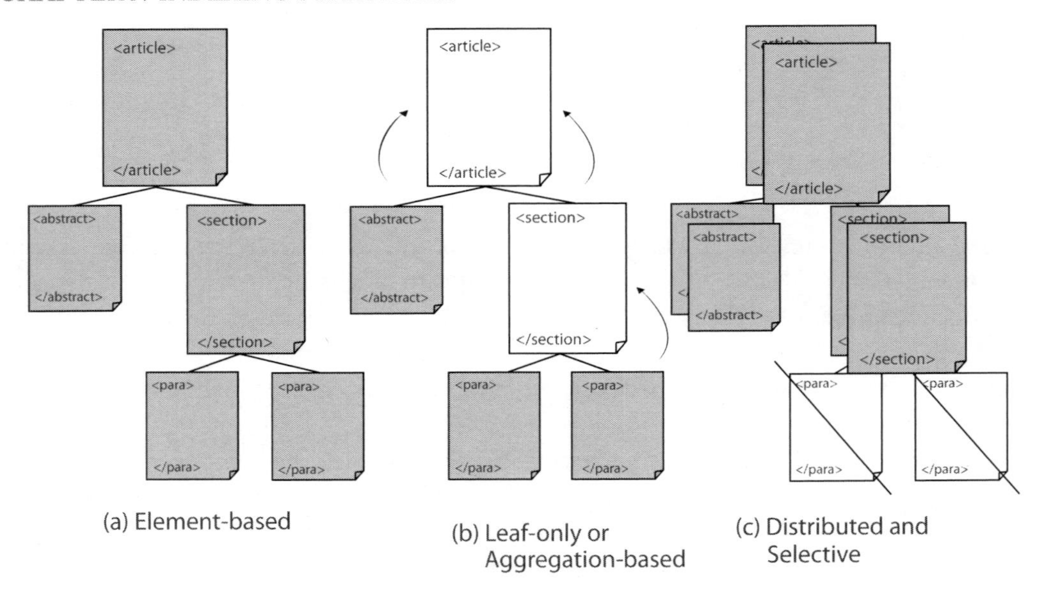

Figure 5.1: Illustrations of some of the indexing strategies, showing in dark the elements in a document being indexed

is composed of two paragraph elements. The fact that a term appears in the paragraph implies that it also appears in the section, which should be taken into account when using ief to discriminate between elements. As a result, alternative ief estimations have been proposed.

In this chapter, we describe indexing strategies developed for XML retrieval. We describe, for each strategy, the choice of elements to index, and the calculation of ief. The two are not unrelated, as deciding which elements to index can affect how ief is calculated. There are five main indexing strategies, namely, element-based (Section 5.1), leaf-only (Section 5.2), aggregation-based (Section 5.3), selective (Section 5.4), and distributed (Section 5.5). Although we describe each strategy separately, various mixtures of strategies have usually been applied. We also cover the indexing of the structure (Section 5.6).

5.1 ELEMENT-BASED INDEXING

The simplest approach to allow the retrieval of elements at any level of granularity is to index all elements. This is illustrated in Figure 5.1(a), where the five elements forming the document, namely, `article`, `abstract`, `section` and `paragraph` twice, are all indexed. Here, each element is indexed based on both the text directly contained in it, and the text of its descendants. For instance, in our example, `section` is indexed on the basis of its own text (if any), and that of its two `paragraph` children elements.

This approach is known as the element-based indexing approach, and is the closest to traditional information retrieval. Each XML element is a "bag of words" of itself and its descendants, and its relevance can be estimated as ordinary flat document. The difference is that the structure is used to decompose documents into retrievable units. A main drawback is that a highly redundant index is generated; indeed, text occurring at level n of the XML logical structure is indexed n times.

Regarding the calculation of term statistics, the most straightforward approach, which directly follows from flat document retrieval, is to compute the term statistics (etf and ief) based on the concatenation of the text of the element and that of its descendants. As already pointed out earlier, this approach ignores the issue of nested elements. As a result, the ief value of a term will consider both the element that contains that term and all elements that do so in virtue of being ancestor of that element (e.g., [165]).

Alternative approaches have, therefore, been proposed. One estimates ief across elements of the same type (e.g., [174]) and another does so across documents (e.g., [29]). The former greatly reduces the impact of nested elements on the ief value of a term, but does not eliminate it as elements of the same type can be nested within each other (as is the case with the Wikipedia test collection used at INEX). The latter is the same as using inverse document frequency, which completely eliminates nested elements.

Experimental results reported in [149] indicate that estimating ief across documents shows slight improvement in results relative to an estimate of ief based on elements within the language modeling framework. Other experimental results produced by a BM25 ranking adapted to XML retrieval [22] suggest that better performance is obtainable by estimating ief across all elements[2] instead of across elements of the same types. At this stage, it is not clear what is the best way to estimate ief, and whether the issue of nested elements in fact matters.

5.2 LEAF-ONLY INDEXING

The element-based indexing strategy leads to a highly redundant index. It also leads to a very large index because the number of elements (potential retrieval units) is much larger than the number of documents. For example, the INEX 2002-2004 document collection is made of approximately 12,000 articles, with a total of 8 millions elements (not counting that many of them are nested!).

One way to reduce the size of the index, as well as removing redundancy, is to index leaf elements only, hence the name leaf-only indexing. This is illustrated in Figure 5.1(b), where the elements shown in dark (the three leaf elements, the `abstract` element and two `paragraph` elements) are indexed. The index built from the leaf elements is used to estimate the relevance of the leaf elements themselves.

A propagation mechanism is then employed to estimate the relevance of non-leaf elements, where the retrieval score of the children elements are combined to form the retrieval score of the parent element [58] (the score combination is described in Section 6.3). The propagation of retrieval scores begins with leaf elements and moves upward the document hierarchical structure. Going

[2] For efficiency issue, not all elements were considered, but a large subset of them.

back to Figure 5.1(b), the propagation is needed to calculate the relevance of the `article` and the `section` elements in our example.

The leaf-only indexing strategy overcomes the issue of nested elements since ief is calculated across leaf elements [160]. It also leads to a more manageable index. However, it requires efficient propagation algorithms for the retrieval of non-leaf elements, which is performed at query time. Indeed, for an article containing thousands of XML elements, a considerable amount of retrieval score propagation may be needed along the article logical structure, in particular to estimate the relevance of elements at the top of the hierarchy (although only some of the document paths – those ending with retrieved leaf elements – are explored).

5.3 AGGREGATION-BASED INDEXING

Instead of using the concatenated text in an element to estimate its term statistics (as in the element-based indexing strategy), another method has been to aggregate the term statistics of the element's own text (if any), with the statistics of each of its children elements (e.g., [64, 145]). The element's obtained aggregated representation is then used to estimate that element relevance (see Section 6.4). The process, if applied at query time (i.e. only leaf elements containing query words and their ancestors are considered), also reduces the size of the index, as only the leaf elements are indexed. This is illustrated in Figure 5.1(b). Here, the aggregated representations of the `section` and then the `article` elements are only computed if at least one of the leaf elements contains some of the query words (are considered relevant).

As for the leaf-only indexing strategy, this indexing strategy can also overcome the issue of nested elements. In addition, the aggregation can include additional parameters incorporating, for instance, structural relationships. Indeed, in a collection of scientific articles, abstracts are likely to contain more informative terms than conclusions. Therefore, the terms in the abstract should have a stronger influence on the terms statistics at the article level, than terms from the conclusions. An important issue is the estimation of the parameters, e.g. reflecting the degree of influence.

5.4 SELECTIVE INDEXING

Although there are no fixed pre-defined retrieval units, not all elements in a document are desirable retrieval units. It is indeed very likely that only a subset of elements constitute meaningful retrieval units. Therefore, a common strategy is to select a subset of elements to be indexed, hence the name selective indexing. This approach has the advantage to reduce the size of the index (although to a lesser extent than the leaf-only indexing strategy).

This indexing strategy is mostly used in combination with another, and as such, the way the terms statistics are computed is usually based on the other strategy. For example, when combined with the element-based indexing, the same concerns (and corresponding estimations) regarding the calculation of ief apply.

Two main selective indexing strategies have been proposed. First, it has been common to discard small elements. In most XML retrieval scenarios, elements made of two to three words are unlikely to be meaningful to users. This selection can be implemented by simply discarding elements smaller than a given threshold (usually expressed in terms of number of words) [165]. It has, however, been shown, that, although small elements may not be meaningful and thus should not be returned, they might still influence the scoring of enclosing elements. Therefore, they may still be indexed, in particular when, as is the case with the leaf-only indexing strategy, a propagation mechanism for scoring non-leaf elements is used [160].

A different approach is to index only elements of selected types (i.e., selected tag names). This is illustrated in Figure 5.1(c), where only `article`, `abstract` and `section` elements are indexed, but not `paragraph` elements. Here, the selection is with respect to the types of elements to be indexed. An alternative, shown in Figure 5.2, is to index so-called indexing units, which consist of disjoint fragments of the documents [50]. The dotted line demarcates each indexing unit, and only fragment of selected types form the roots of index units, in our case `book`, `chapter` and `section`. The reason to have disjoint fragments was to overcome the issue of nested elements when calculating term statistics[3].

The selection can be based on the analysis of available assessment data. For instance, let us assume that for a test collection, such as the IEEE test collection used at INEX, the most frequently judged relevant elements are of types paragraph, section, subsection, and article. The outcome would be then to only index elements of these types, which is the approach adopted in [127]. The assessment data could also consist of logs data, from where the most clicked element types can be extracted. This approach, however, requires a working XML retrieval system that logs this information. Finally, the selection of the types can be manual. For example, the system designer could choose the types of elements to be indexed based on the analysis of the DTD of the document collection, the application and user requirements.

5.5 DISTRIBUTED INDEXING

In this strategy, a separate index is built for each element type. The term statistics for each index are then calculated separately. Since each index is composed of terms contained in elements of the same type, it will be made of entries, which in terms of vocabulary and size, are likely to be more uniform than if coming from an index built from elements across all types. This is likely to result in more consistent term statistics. Indeed, ief become calculated across elements of the same type. In addition, because of this, the distributed indexing strategy greatly reduces the term statistics issue arising from nested elements, although it does not eliminate it (elements of the same type can be nested). At retrieval time, the query is then ran in parallel on each index, and the lists of results (one for each index) are merged to provide a single list of results, hence the name distributed indexing. We discuss the merging operation in Section 6.5.

[3]The relevance of an indexing unit, for example, the second `chapter` in Figure 5.2, is estimated through a propagation (referred to as augmentation in [64]) mechanism, so that to consider all that unit content, e.g. here the two `section` elements.

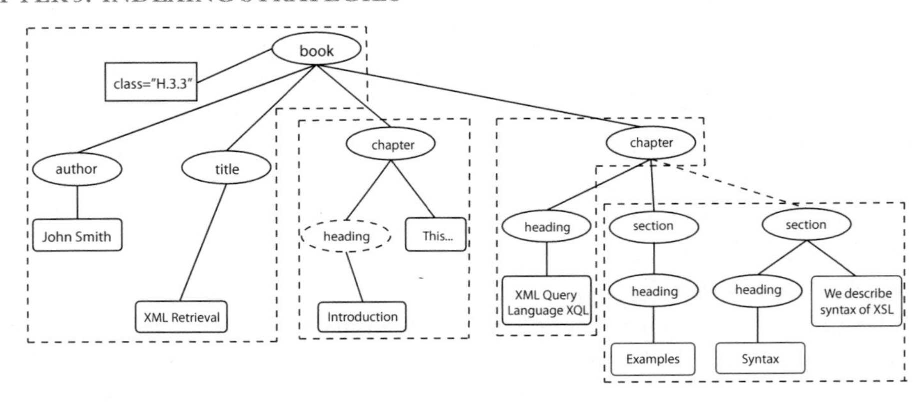

Figure 5.2: Disjoint indexing unit selective strategy (taken from [50])

This indexing strategy has been used in conjunction with the selective indexing strategy. This is illustrated in Figure 5.1(c), where an `article` index, an `abstract` index and a `section` index are built, but no `paragraph` index. Indeed, it will often not make sense to build an index for all types of elements, as not all of them are meaningful retrieval units. This is the approach taken by Mass and Mandelbrod [127], who actually proposed the distributed indexing strategy. In their work, the selected types (this is actually the selective indexing discussed in the previous section) included article, abstract, section, subsection, and paragraph, which corresponded to the type of elements with the highest distribution of relevant elements in the collection. These element types are, indeed, particularly meaningful for content-oriented XML retrieval.

5.6 STRUCTURE INDEXING

It may be beneficial to distinguish different "structural contexts" of a term when computing term statistics. Indeed, the fact that a term appears in one element may have nothing to do with the same term appearing in another element. To distinguish the structural context of a term, one approach is to compute statistics for structure/term pairs. The structure can be the actual path, from the root of the document to the element itself. Let us assume that the term "xml" appears in two elements, with respective path "/article/body/section" and "/article/abstract". The idea is then to construct two structured terms, one for each path; in our example, these would be "/article/body/section(xml)" and "/article/abstract(xml)" (the syntax is not important here). Standard term statistics could then be applied on these structured terms, e.g. [163].

Retrieval will be based on matching these structured terms to those of the query. It is important to allow for an approximate matching of the structural contexts, e.g. [111, 7], otherwise the process would lead to too few matches. Unfortunately, this approach runs into sparse data problems; indeed, many structure/term pairs occur too rarely to reliably estimate meaningful statistics. A compromise is to consider the element tag only, forming tag/term pairs, for instance, "section(xml)" and "abstract(xml)".

Considering the structural contexts has been used differently than through the construction of structure/term pairs. The idea is to associate a weight to a structural context to reflect its significance. These weights are then used in the scoring function used to estimate an element relevance. Several ways to calculate these weights exist. The weights have been based for example on depth (of the path) and location (e.g. "article/section[1]" vs. "article/section[2]") in the document logical structure, and then used as prior probabilities in a retrieval function based on the language modeling framework [73]. The weights have also been calculated based on the distribution of the structural contexts in the collection e.g. [99]. When the structural context is restricted to the type of the element (its tag), an interesting approach is that of Gery et al. [57], in which the distribution of tag names is used in a way similar to the binary independence retrieval model, but here, investigating the "presence" of tags in relevant and non-relevant elements, to estimate the tag weights. These weights can then be used, for instance, in a BM25 scoring function.

5.7 DISCUSSION

In this chapter, we described indexing strategies developed in XML retrieval. We also discuss how the term statistics are calculated within these strategies. Indexing strategies have often been used together, in particular the selective indexing strategy in tandem with another strategy.

The element-based indexing approach uses the document logical (hierarchical) structure to decompose each XML document into retrieval units. The decomposition ignores other structural characteristics of the document, for example, the element types (unless combined with the selective indexing strategy). The element-based indexing does not require access to the collection DTD (or XML Schema). This may be important as not all collections have an associated DTD (as is the case with the Wikipedia document collection used at INEX).

With the selective indexing approach, the specific tags, their semantics and/or importance can be taken into account when selecting the elements to index. Knowing which are the more important elements or element types (e.g., from relevance data) can be powerfully exploited. The drawback, however, is that the choice of elements or element types, and thereby the effectiveness of the approach, is collection-, application- and/or scenario-dependent.

The element-based indexing approach builds a highly redundant index. This may not be as undesirable as it may appear, since the redundant index has essentially "precomputed" term statistics per element, that otherwise need to be computed at query time. In contrast, the leaf-element indexing strategy requires much less storage space. However, elements at the top of the document hierarchy require considerable propagation of retrieval scores, which is performed at query time. However, the propagation will not involve all document paths, only those ending with relevant leaf elements, thus not likely to impede much (if at all) on efficiency, compared to the element-based indexing.

Although both start from an index of leaf elements, the propagation mechanism used with the leaf-only indexing strategy is different to the aggregation-based indexing strategy. With the former, retrieval scores are combined, whereas with the latter, term statistics are combined. This can be compared to an outer (propagation) vs. an inner (aggregation) combination mechanisms. This can be related to the work of Robertson et al. [152], where a similar difference is investigated in field-based retrieval. It was found that the inner combination was better (in terms of retrieval effectiveness). It would be interesting to compare the two types of combination in the context of XML retrieval.

It is not yet clear which indexing strategy is the best, as obviously which approach to follow would depend on the collection, the types of elements (i.e., the DTD) and their relationships. Also mixing strategies has been beneficial. In addition, the choice of the indexing strategy (or a mixture of them) has an effect on the ranking strategy. An interesting research would be to investigate all indexing strategies within a uniform and controllable environment to determine, through both experimental and theoretical analysis, those leading to the best performance for various and even across ranking strategies.

The next chapter describes a second core task of an XML retrieval system, namely that of estimating an element relevance, in order to produce a ranked list of results, here XML elements, as answers to a query.

CHAPTER 6

Ranking strategies

After indexing, a second main task of an XML retrieval system is to determine those elements in the collection that are relevant to a given query. These elements are then presented ranked according to their estimated "degree" of relevance (but see Chapter 7, where alternative presentation strategies are described). In classical information retrieval, estimating the relevance of a document to a query is done through information retrieval models [124]. Many information retrieval models exist in the literature, for example, the Boolean, the vector space (e.g. tf $\times$ idf) and probabilistic models (e.g. language models).

Many of the retrieval models developed for classical information retrieval, i.e. "flat" document retrieval, have been adapted to XML retrieval. These models have been used to estimate the relevance of an element based on the evidence coming from the content of the element only (Section 6.1). It has often been the case that other evidence has been used to estimate an element relevance. Indeed, in particular for long documents, using evidences coming from the context of the element (e.g., the parent element) has been shown to be beneficial. This strategy is referred to as contextualization (Section 6.2).

As discussed in Chapter 5, several indexing strategies have been developed in XML retrieval. Depending on the chosen strategy, specific mechanisms, such as propagation (Section 6.3) and aggregation (Section 6.4) – when only leaf elements are indexed, and merging (Section 6.5) – when distributed indexing is used, are needed to rank elements at all levels of granularity.

The ranking strategies described from Section 6.1 to Section 6.5 are concerned with estimating the relevance of elements (at all level of granularity) with respect to a content criterion. For a content-and-structure query, additional processing is necessary to provide a ranked list of elements that not only satisfy the content but also the structural criteria of the query (Section 6.6).

6.1 ELEMENT SCORING

All ranking strategies require a scoring function that estimates the relevance of an element to a given query. With the propagation strategy (discussed later in this chapter), the scoring function is applied to the leaf elements only, whereas in other cases, it is applied to all retrievable elements. The scoring function is based on standard information retrieval models, such as the vector space, BM25 and language models, which are adapted to XML retrieval. As an illustration, we describe a scoring function based on a language modeling framework [146]. This illustration is inspired by the work of Sigurbjörnsson et al [165].

Given a query $q = (t_1, t_2, ..., t_n)$ made of n terms t_i, an element e and its corresponding element language model M_e, elements are ranked in decreasing order of $P(e|q)$, which is expressed as follows[1]:

$$P(e|q) \propto P(e)P(q|M_e)$$

where $P(e)$ is the prior probability of relevance for element e and $P(q|M_e)$ is the probability of the query q being generated by the element language model M_e. Using, for instance, a multinomial language model with Jelinek-Mercer smoothing [105], $P(q|M_e)$ is calculated as:

$$P(t_1, t_2, ..., t_n|M_e) = \prod_{i=1}^{n} \lambda P(t_i|e) + (1 - \lambda)P(t_i|C)$$

where $P(t_i|e)$ is the probability of a query term t_i in the element e, $P(t_i|C)$ is the probability of a query term t_i in the collection, and λ is the smoothing parameter.

To account for the wide range in element sizes in XML documents (e.g., from paragraph to article), and in particular for the heavily biased distribution of small elements in XML document collections (e.g., there are many more paragraphs than sections), a bias towards long elements can be incorporated by setting $P(e)$ as follows[2]:

$$P(e) = \frac{length(e)}{\sum_{e'} length(e')}$$

where the summation in the denominator is with respect to all indexed elements (the set of indexed elements depends on the indexing strategy used, see Chapter 5).

Other estimates of prior probabilities include the number of topic shifts in an element [13] and the path length of an element [73] (the latter corresponds to the structure indexing - see Section 5.6). Overall, considering element length has been shown crucial in element retrieval, as this caters for the wide range in element sizes, whereas other estimates of the prior probabilities can help, depending on the retrieval task [13].

The scoring function, as illustrated here, provides an estimate of the element relevance, i.e. a retrieval score, for a given query. However, the generated scores are not necessarily used or cannot be used as such to produce the ranked list of elements for the query. Additional strategies have often been used or are needed to produce the final ranked list of elements.

6.2 CONTEXTUALIZATION

The terms found in some types of elements will often form a subset of those used in other types. For instance, for the INEX 2002-2004 document collection, the paragraph index has 25% less entries than the article index [127]. Taking into account this variation in term vocabulary can be beneficial for XML retrieval, in particular when estimating the relevance of elements down in the document hierarchy, which tend to be small. This can be done by considering the context of an element to

[1]Throughout this chapter, simplified versions of the formulae are given.
[2]In [78], a log prior is used.

estimate the relevance of that element. Context here refers to the parent of the element, all or some of its ancestors, the whole document (the root element).

The context of an element can provide more evidence on what an element is or is not about. This is because all the terms in the collection, whether in the element or not, can be used to score (estimate the relevance of) the element for a given query. For instance, the fact that an element does not contain all query terms, but is contained in a document that contains all query terms, is likely to be more relevant than if contained in a document that does not contain all query terms. This strategy can be implemented by combining the element retrieval score to that of the document score.

The process of combining the retrieval score of the element and that of its context is referred to as contextualization. The most common contextualization technique is indeed to use the document containing the element (i.e. the root element) as context. This means combining the score of the element to that of the XML document containing that element, where the element and the document retrieval scores is estimated for example as described in Section 6.1. The combination can be as simple as the average of the two scores [10]. A scaling factor can be used to emphasize the importance of one score compared to the other [127]. This contextualization technique (using element and document scores) has been shown to increase retrieval performance.

Other contextualization techniques have been used. Some or all ancestors of an element can also be used as context. For instance, in [10], the parent element alone is used as context. All the ancestor elements are also used as context. The various scores are summed for all the elements considered in the contextualization process. Experiment results show clear improvement over not using any context, where the best results were obtained using all ancestors as context.

In [192], the element score, that of its parent, and that of its root (the whole document), are linearly combined, and machine learning techniques are used to estimate the parameters (the scaling factors) for the combination. Here, BM25 was used as the scoring function. It was also shown that this combination outperforms using the element score only.

A main issue when combining several contexts is that it is necessary to scale the effect of the different contexts, which may require training data and extensive experimentation. Nonetheless, the contextualization strategy has been shown to be successful compared to using the element score alone to rank elements, in particular for long documents.

6.3 PROPAGATION

The propagation strategy is needed with the indexing strategy (see Section 5.2) that only indexes leaf elements. The relevance of the leaf elements for given queries is estimated on this indexing, resulting in retrieval scores for leaf elements. The relevance of non-leaf elements (the inner elements, including the root element) is estimated through a propagation mechanism, where the retrieval score of an inner element is calculated on the basis of the retrieval scores of its descendant elements. The propagation starts from the leaf elements and moves upward in the document tree structure. The most common propagation mechanism consists of a weighted sum of the retrieval scores, where the variations come with the definition and estimation of the weights.

Let e be a non-leaf (inner) element, and e_l a leaf element of e[3]. Let q be a query. Let $score(.)$ be the scoring function used to rank elements according to how relevant they have been estimated. For e_l, $score(e_l, q)$ is calculated directly from the index of e_l (e.g. using an element-only scoring strategy as described in Section 6.1), whereas $score(e, q)$ is calculated through the propagation mechanism.

The distance, implemented as the length of the path between an inner element and its leaf elements, has been used as a weight. The idea here is that the greater the distance between an element and its leaf elements, the less the retrieval scores of the leaf elements should contribute to that of the inner element. In addition, for normalization purpose, the distance between the root element, denoted below $root$, and the leaf elements has also been used as weight in the summation. The following scoring function illustrates the use of these two distances as weights [74]:

$$score(e, q) = \sum_{e_l} (1 - 2.\lambda.\frac{d(e, e_l)}{d(e, e_l) + d(root, e_l)})^2 \times score(e_l, q)$$

where λ is a constant and $d(x, y)$ is the distance between elements x and y in the document tree. The above propagation mechanism views an element e that has only one retrieved descendant element as being less relevant, whereas an element e that has several retrieved descendants is viewed as more relevant than any of the descendants. Near descendants have a stronger impact than those farther away. Finally λ allows varying levels of contribution of the descendants in the propagation, where $\lambda = 0$ means that $score(e, q)$ is simply the sum of the scores of all the retrieved leaf elements.

The number of children elements of an element has also been used as a weight. For instance, in the GPX approach [58], $score(e, q)$ is calculated as follows[4]:

$$score(e, q) = D(m) \sum_{e_c} score(e_c, q)$$

where e_c is a child element of e, m is the number of retrieved children elements of e, $D(m) = 0.49$ if $m = 1$ (e has only one retrieved child element), and 0.99 otherwise. The value of $D(m)$, called the decay factor, depends on the number of retrieved children elements. If e has one retrieved child, then the decay factor of 0.49 means that an element with only one retrieved child will be ranked lower than its child. If e has several retrieved children, the decay factor of 0.99 means that an element with many retrieved children will be ranked higher than its children elements. Thus, a section with a single relevant paragraph would be considered less relevant than the paragraph itself (as it is simply better to return the paragraph as returning the section does not bring anything more), but a section with several retrieved paragraphs will be ranked higher than any of the paragraphs (as it will allow users to access these several paragraphs through the returned section). Although the idea behind GPX is comparable to the approach previously discussed, GPX led to better results.

[3]There is path between e and e_l, where e_l is the last element of the path, i.e. e_l has no children elements.

[4]When the weight reflects the number of children, the weight will be the same for all children elements, and is thus outside the summation.

We finish with the XFIRM system [160], where various weights are used in the propagation mechanism, together with the contextualization strategy. The (simplified) score of an inner element e for a query q is given as follows:

$$\begin{aligned} score(e,q) &= \rho \times m \times \sum_{e_l} \alpha^{d(e,e_l)-1} \times \beta(e_l) \times score(e_l,q) \\ &\quad +(1-\rho) \times score(root,q) \end{aligned}$$

where m is the total number of retrieved leaf elements contained in e (in other words, those e_l for which $score(e_l,q) > 0$). $score(root,q)$ is the retrieval score of the $root$ element. Here, contextualization is applied by adding the document score ($score(root,q)$) to the propagated element score (as calculated in the summation) to form the overall element score ($score(e,q)$), where ρ is a parameter that is used to emphasize the importance of the propagated element score versus that of the document score. $\beta(e_l)$ is used to capture the importance of leaf elements in the propagation. These elements may correspond to small elements, such as emphasized words (e.g. bold, italic) or titles, and as such can be viewed as containing important terms. For example, a high β value can be assigned to title leaf elements compared to any other leaf element types, to increase the impact of retrieved titles in the propagation (i.e. as important terms are matching the query terms). This explains the need, as advocated by Sauvagnat et al. [160], to index small elements, even if they do not constitute retrieval units (see Section 5.4).

The propagation mechanisms described in this section led in general to good retrieval performance, in particular that implemented by the GPX system. Although rather simple, the GPX propagation mechanism produced top performance across years and retrieval tasks at the INEX campaign, thus showing its versatility for XML retrieval.

6.4 AGGREGATION

The aggregation is based on the work of Chiaramella et al on structured document retrieval [27]. The basic idea is that the representation of an XML element (a structured component) can be viewed as the aggregation of its own content representation (if any) and the content representations of structurally related elements (if any). The common practice in XML retrieval has been to perform the aggregation following the document hierarchical structure, which means that the aggregation is with respect to the element own content (again if any) and that of its children elements. Retrieval is then based on these aggregated representations. Aggregation is to be contrasted to propagation; in the former, the combination is applied to representations, whereas in the latter, it is applied to retrieval scores.

The element own content representation is generated using standard indexing techniques, whereas an aggregation function is used to generate the representation of the non-leaf elements. The aggregation function can include parameters (referred to as augmentation factor in [64] or accessibility weight in [154]) specifying how the representation of an element is influenced by that of its children elements (a measure of the contribution of, for instance, a section to its embedding chapter).

The aggregation process can take place either at indexing time (global aggregation) (e.g. [99]) or at query time (local aggregation) (e.g. [145]). Global aggregation, which considers all indexing terms, does not scale well and can quickly become highly inefficient. As a result, local aggregation strategies are primarily used, where the aggregation is restricted to query terms. It starts from elements retrieved in terms of their own content, and then the aggregation is performed only for the ancestors of these elements.

To illustrate the aggregation mechanism in XML retrieval, we describe an approach based on the (simplified) language modeling framework, which is inspired from the work of Ogilvie and Callan [145]. In this framework, each element is modeled by a language model. For an element e, the probability of a query term t_i given a language model based on the element *own* content $M_{e_{own}}$ is given by:

$$P(t_i|M_{e_{own}}) = (1-\lambda)P(t_i|e_{own}) + \lambda P(t_i|C)$$

where λ is the smoothing parameter.

Now assume that e has, in addition to its own content, several children, e_j, each with their own language model M_{e_j}. Then, the aggregation function can be implemented as a linear interpolation of language models:

$$P(t_i|M_e) = \omega_0 P(t_i|M_{e_{own}}) + \sum_j \omega_j P(t_i|M_{e_j})$$

where $\omega_0 + \sum_j \omega_j = 1$. The ω parameters model the contribution of each language model (i.e., child element, own content) to the aggregation.

The ranking of the elements is then produced by estimating the probability that each element generates the query, which can be defined as follows, where the query term is made of n terms, t_1 to t_n:

$$P(t_1, t_2, \ldots, t_n|M_e) = \prod_{i=1}^{n} \lambda P(t_i|M_e)$$

Other approaches for dealing with aggregation use fielded BM25 [119] and probabilistic models [64, 99]. Earlier approaches (not evaluated in the context of INEX) include [112] and [132]. An important issue with the aggregation method is the estimation of parameters (e.g., the estimation of the ω values above).

6.5 MERGING

Some approaches developed for XML retrieval produce separate ranked lists of results for a given query, which are then merged to return to the user a single ranked list of results [19, 116, 127].

The approach described in [127] uses a distributed indexing strategy (Section 5.5), in which a separate index is created for each element type (e.g., for a collection of scientific articles, these include article, abstract, section, paragraph). Let us now assume that a retrieval model is used to rank the elements in each index (in [127], the vector space model is used). This results in separate ranked lists, one for each index. To merge the lists, normalization is necessary to take into account the variation in size of the elements in the different indices (e.g., paragraph index vs article index). For this purpose, for each index, $score(q, q)$ is calculated, which is the score of the query as if it were an element in the collection. For each index, the element score is normalized with $score(q, q)$ so that any element identical to the query obtains a full score of 1. This ensures that scores across indices are comparable, and that the elements (retrieved from separate indices) can then be merged based on the normalized scores.

Amati et al [4] also use a distributed indexing approach, but in addition, use several statistical retrieval models to rank elements, and normalization functions, including the one above described. They show that an appropriate choice of both the ranking model and the normalization can have a positive effect on retrieval performance.

A second approach where merging is performed is when several ranked lists of results are generated for all elements in the collection (as opposed to separate indices, as above described). The difference comes with the use of several retrieval strategies (e.g. ranking models, processing of the query), each producing a list. This can be compared to data fusion investigated in the mid 90s (e.g. [194]). Examples of such merging approaches for XML retrieval include [116, 19].

6.6 PROCESSING STRUCTURAL CONSTRAINTS

The ranking strategies described so far in this chapter were concerned with estimating the relevance of an element according to how the element content matches the query terms. This is exactly what is needed for content-only queries (Section 4.2.1). As discussed in Chapter 4, and more precisely in Section 4.2.2, with XML documents came content-and-structure queries, which allowed expressing constraints with respect to the content and the structure of the elements to be retrieved. Examples of XML languages were provided in Chapter 4. In this section, we describe approaches that were developed to process structural constraints expressed within NEXI, the path-based query language used at INEX (Section 4.3.2).

In INEX, structural constraints are viewed as hints as to where to look to find relevant information. The reason for this view is two-fold. First, it is well known that users of information retrieval systems do not always or simply cannot properly express the content criterion (i.e. select the most useful query terms) of their information need. We recall that with these systems, queries are with respect to content only. It is very likely that this difficulty also holds for the structural criterion of the information need. For instance, a user asking for paragraph components on some given topic may not have realized that relevant content for that topic is scattered across several paragraphs, all of which contained within a single section. It would thus be more useful to return that section instead of individual paragraphs.

Second – and to some extent as an indirect consequence of the first reason above – there is a strong belief in the XML (information retrieval) community that satisfying the content criterion is, in general, more important that satisfying the structural criterion. For instance, even if a user is looking for section components on a particular topic, returning to that user abstract components would still be satisfactory, as long as the content criteria is "well" satisfied. A number of approaches have been developed to process structural constraints in XML retrieval following this view.

A first approach is to build a dictionary of tag synonyms. The dictionary can be syntactically based. If, for example, `<p>` corresponds to paragraph type and `<p1>` corresponds to the first paragraph in a sequence of paragraphs, it would be quite logical to consider `<p>` and `<p1>` as equivalent tags (e.g. [126, 159]). It can be semantically based, for instance, considering `<city>` and `<town>` as equivalent tags. The dictionary can also be built from processing past relevance data [130]. If in such a data set, for example, a query asked for `<section>` elements, then all types of elements assessed relevant for that query are considered equivalent to the `<section>` tag. Thus with this approach, if the structural constraint refers to e.g. `<section>`, then there is no distinction made between section elements and elements whose type is synonymous with`<section>` in the dictionary. These elements would then be returned as answers if the content criterion is satisfied.

A second technique is that of structure boosting. There, the retrieval score of an element is generated ignoring the structural constraint of the query, but is then boosted according to how the structural constraint is satisfied by the element. The element structure and the query structure are compared and some structure score is generated depending on the level of vagueness. For instance, the paths can be compared (e.g. [174, 25], or the tags in the paths are compared [190]. Structure boosting is also used when the propagation method is used to score (non-leaf) elements. When propagating the retrieval scores from children to parent elements, it is possible to boost the resulting score when the parent element matches the structural constraint [74]. An important issue here is what is the actual level of vagueness (how much of a hint is the structural constraint). A flexible approach to capture various levels of vagueness in the ranking function was proposed in [111].

Consider the query "retrieve paragraphs about ranking algorithms contained in sections about XML retrieval". Processing this query involves a first step, which is to divide the query into two sub-queries, e.g. "retrieve paragraphs about ranking algorithms" and "retrieve sections about XML retrieval". Each sub-query is then processed as described above, e.g. using a tag synonym dictionary or through structure boosting (here with respect to matching tags). Each sub-query results in a ranked list of elements. To generate a ranked list for the whole query, the two ranked lists are compared, e.g. only elements returned for the "paragraph" sub-query whose ancestors are "contained" among the elements returned for the "section" sub-query are then retrieved. The final score depends on the implementation of "being contained", e.g. strict containment or fuzzy containment (e.g. [193]).

Although not described here, in database research, a wealth of approaches have been proposed to vaguely match the structural constraints, referred to as query relaxation (e.g., [7]). They mainly consisted of matching the document tree structure to the query tree structure. It should be noted that the queries were not expressed in NEXI. In addition, these approaches were not evaluated in terms of effectiveness, or at least not in a large scale (e.g. using a test collection such as that provided by INEX).

The techniques described to process content-and-structure queries were evaluated in the context of INEX where the relevance of an element was assessed based on content only (see Section 8.3). This is because, as described at the beginning of this section, for INEX, satisfying the content criterion of a query is what really matters, and the structural constraints of the query should simply be viewed as hints about, for example, the types of elements where relevant content can usually be found for that query. However, when looking closely at the actual structural constraints, it was found by Trotman and Lalmas [185] that they were not really hints. Although a query asked for a section, being a section element was not really a structural hint. In other words, there was no indication from the relevance data set that a section element was a better element type to return than another element type. Finally, and likely as a consequence of this, using the structural constraints in the query did not usually increase retrieval performance, compared to using the content constraints only (see Section 8.2 for details about this), apart for maybe at very early ranks.

These findings are rather disappointing and it is important to re-visit the evaluation methodology adopted by INEX before any new insights can be made regarding the usefulness of structural constraints in XML retrieval. The disappointing results also likely due to the collections used at INEX, as the tags in both the IEEE and the Wikipedia collections (see Section 8.1) relate to section, paragraph, abstract, and so on, many of which being rather general. It would be interesting to observe the effect of structure-and-content queries for patent retrieval where the text occurring in the "claims" component of a patent document is very different to the text forming the "prior art" component.

6.7 DISCUSSION

In this chapter, we described strategies used for ranking XML elements. It is not yet possible to unanimously state which ranking strategy works best since many factors are involved when deciding how relevant an element is to a given query (e.g. the size of an element, the type of element, the relevance of structurally related elements, etc). Indeed, XML retrieval can be regarded as a combination problem where the aim is to decide which evidence to combine and how to combine it for effective retrieval. We can, however, postulate that including the context of the element in addition to the element own content (directly or using a propagation or aggregation strategy) to estimate that element relevance to a given query has been shown to be beneficial for XML retrieval. An important issue is the appropriate estimation of the parameters used in combining the evidence, which is crucial for effective retrieval. A remaining issue is the processing of structural constraints in content-and-structure queries that could lead to improved retrieval performance.

The ranking strategies described in this chapter have been developed with the purpose of estimating the relevance of an element for a given query. This is often not the end task in XML retrieval. Indeed, it may not make sense to return all (estimated) relevant elements, as elements may have relationships (a section element and its paragraph element). Deciding which of the relevant elements to return and how to present these selected elements are discussed in the next chapter.

CHAPTER 7

Presentation strategies

In the previous chapter, we described approaches developed to estimate the relevance of an XML element to a given query, for content-only and content-and-structure queries. This estimate results in a retrieval score, which is then used to produce a list of elements, ordered according to their calculated relevance. Figure 7.1 shows an interface displaying such a ranked list of XML elements. This, however, may not be the most useful way to return results in XML retrieval. This is mainly because XML elements are not independent, e.g. two retrieved elements may overlap (e.g. a section element and one of its paragraphs) or may be siblings (e.g. two consecutive paragraphs of a section), which may need to be taken into account when presented to users.

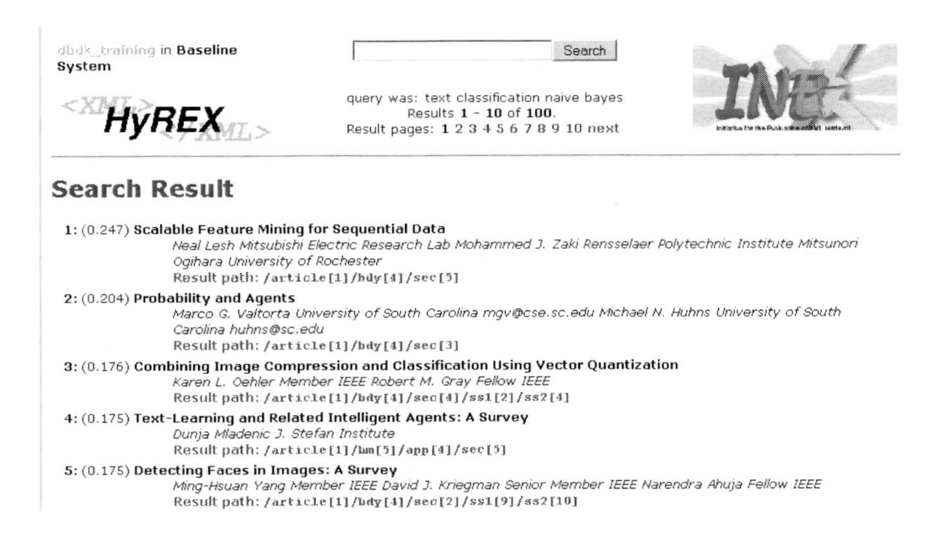

Figure 7.1: Presenting a ranked list of elements (reproduced from [64])

Since 2004, INEX runs an interactive track (iTrack) that looked at interaction issues in XML retrieval [176], and in particular the impact of result dependency on user satisfaction. One outcome of iTrack is that users did not like being returned with (at least too much) redundant information (overlapping results). This led to the development of approaches specifically dedicated to generate a list of results that has no or little overlaps (Section 7.1). A second outcome was that users expected to have not only access to relevant elements, but also to their context (e.g. documents containing the retrieved elements, sibling elements). This led to various proposals of presenting XML retrieval results in context (Section 7.2). Other studies (e.g. [151]) showed the usefulness of returning entry points to a document, instead of elements. This led to the development of approaches aimed at identifying the best entries points in documents (Section 7.3).

Dealing with overlaps, presenting results in context and determining best entry points are retrieval tasks that have been and are still being investigated at INEX, namely, the focused task, the relevant in context task, and the best in context task, respectively (Section 8.4). INEX has for a while acknowledged that estimating the relevance of an element to a given query is not the only retrieval task in XML retrieval. Therefore, these other retrieval tasks were set up to investigate different ways to present results in XML retrieval. We will be referring to these tasks throughout this chapter (but see Section 8.4, which provides a detailed description and chronology of these tasks).

7.1 DEALING WITH OVERLAPS

We recall that an XML retrieval system aims at returning the most relevant elements for a given user query. When an element has been estimated relevant to a given query (by any of the XML ranking strategies presented in the previous chapter), it is likely that its ancestors will also be estimated relevant (although likely to a different extent). Furthermore, this element most probably contains several descendant elements that have also been estimated as relevant (also likely to a different extent). This is due to the nested structure of XML documents where the same text fragment can be contained in several elements along a same path. Thus the element itself, its ancestors and a number of its descendants may be contained in the result list, eventually, leading to a considerable amount of redundant information being returned to users.

Returning redundant information (i.e., overlapping elements) has been shown to distract users [177]. In retrieval scenarios where users do not like or wish to see the same information several times, XML retrieval systems need to decide which of these relevant but overlapping elements should be presented to the users. This retrieval task, which corresponds to the focused retrieval task investigated at INEX (see Section 8.4 in the next chapter), is different to the task of ranking the elements in terms of how relevant they are to a given query.

The simplest way to deal with overlaps is to select a subset of element types and consider only these for retrieval, e.g. considering only section elements and ignoring all the other element types. Although overlap is reduced, it is not completely removed, as it is still possible that elements of the same type overlap each other. In addition, this approach requires a priori knowledge of the collection usage, e.g. what are the most useful element types for all queries. It also lacks flexibility as the most useful element types may vary in granularity depending on the queries. As a consequence, more flexible approaches have proposed, where the aim is to remove (or reduce) overlaps from the ranked list of elements initially produced by an XML retrieval system.

The most commonly adopted approach, referred to as brute-force filtering, selects the highest ranked element from the result list produced by an XML retrieval system and removes any ancestor and descendant elements at lower ranks. The process is then applied iteratively, and relies on the retrieval strategy to rank, among overlapping elements, those that should be selected at higher ranks. However, although the tree structure of a document at indexing and/or retrieval time may have been used to rank elements, it does not means that the ranking is appropriate for the purpose of returning the list of the most relevant but non-overlapping results. For this reason, other approaches to control or remove overlaps in retrieval results have been put forward.

One approach that is concerned with controlling the amount of overlaps is that of Clarke [29]. Here the initial ranking produced by the XML retrieval system is modified according to a tolerated amount of redundant information. The retrieval score of an element that is contained or contains other retrieved elements ranked higher are adjusted to reflect that the information it contains is redundant (i.e. has already been seen). This is done by reducing the importance of terms occurring in already seen elements in the ranking formula. This approach is not designed to remove overlaps, although it can, but to push down the result list those elements that contain redundant information. In addition, it can be tuned depending of the amount of redundant information tolerated by the user or retrieval scenario.

Another technique that re-ranks the initial list of results is that of Popovici et al. [147]. There, the retrieval score of an element is re-calculated with the retrieval scores of its (if any) descendent elements. This is done through a bottom-up propagation mechanism, using for instance, the maximum or average operation to re-calculate the scores. The new ranked list is then filtered by selecting the highest ranked elements, and then removing either all ancestors or all descendants of that selected element from the list. This filtering process is applied iteratively. Thus, differently to the previous approach, an overlap-free result list is produced. Also, differently to the brute force filtering, either ancestors or descendants are removed, not both. The best approach is the one using the maximum function to re-rank elements and removing the descendants of each selected element.

The next two approaches are concerned with presenting an overlap-free result list to users, where the tree structure of the document is explicitly used to select the elements that will form that list, as opposed to the brute-force filtering or a re-ranking of the results (or a combination of both).

The first one, [130], is based on the notion of the usefulness of an element, which is modeled through an utility function based not only on the retrieval score of an element, but its size, and the amount of irrelevant information contained in its children elements (in other words, "amount" of text contained in the non-retrieved children elements). If an element has an estimated utility value higher than the sum of the utility values of its children, then the element is selected and the children are removed. Otherwise, the children elements whose utility values exceed some threshold are selected and the element is removed.

A second approach [128] looks at the distribution of retrieved elements in the XML document tree structure in addition to their score. For instance, an element that has many of its descendants retrieved, but which are evenly distributed in the corresponding tree structure, and in addition has a similar score to the parent element is selected – this is because already from that selected element, all its descendants, many of which being estimated as relevant, can be accessed; otherwise its descendants are selected to be themselves processed.

Overall, techniques that explicitly consider the document tree structure or re-rank elements to remove (or reduce) overlaps outperformed those that do not, i.e. the brute-force filtering approach.

The latter relies on a good initial ranking for the purpose of selecting relevant but non-overlapping elements. There is, however, the issue of speed, as the removal of overlaps is done at query time, thus requiring not only effective but efficient techniques. An interesting question would be to investigate the effect of the original result list (how good it is) on the overlap removal strategy, including the brute-force filtering strategy. There are indications that a good initial result list leads to a better overlap-free result list, than a less good one [12].

7.2 PRESENTING ELEMENTS IN CONTEXT

One finding of the iTrack [177] is that users prefer retrieval results to be presented in context. There were two main criticisms. First, users expected to have not only access to a relevant element, but also to the whereabout of that element, e.g. for a retrieved paragraph, the section and even the document containing that paragraph. Second, users did not like to be presented elements from a same document, for example siblings, scattered across the ranked list of results, as this disrupted their reading flow.

One way to respond to the first criticism is to show a table of contents in addition to the element being viewed by the user. This is illustrated in Figure 7.2. Here, two panels are displayed, one with the element being viewed on the right (a section entitled "Entertainment"), and the second with the table of contents. In that second panel, the title of the element being viewed can be highlighted to indicate where in the document the element is located.

The table of contents can be either fixed or automatically generated. An example of the former is for the table of contents to correspond to the document structure, i.e. the titles of all its sections, subsections, etc. This may, however, not be the most effective way to support users in their information seeking tasks. For instance, studies [170, 171] showed that the table of contents should be query-biased, thus helping users locating other relevant elements. A table of contents should also not be large in size, i.e. longer documents should still have a relatively small table of contents (twenty components was found to be an optimal number in these studies), thus suggesting that an automatic generation of the table of contents has to be more carefully designed when dealing with longer documents (e.g. books). The automatic generation of a table of contents can be compared to the task of producing a summary of the document structure (as opposed to a summary of the document text), one that is element- and query-biased [168].

Addressing the second criticism led to the set-up of the relevant in context retrieval task at INEX [121]. The aim of this task is to return a list of documents, ranked according to their estimated relevance, and within each document, identify the most relevant elements. That is, elements of the same documents are not scattered anymore across the ranking, but always presented together. Figure 7.3 shows an example of an interface that displays the outcome of such a task. Here, a heat-map metaphor is used, where the more relevant the element, the darker the associated color.

Several, although related, approaches were proposed to implement the relevant in context task at INEX. A common approach is to start from an overlap-free ranked list of elements, e.g. as generated by the methods described in Section 7.1 (the focused retrieval task). These elements are

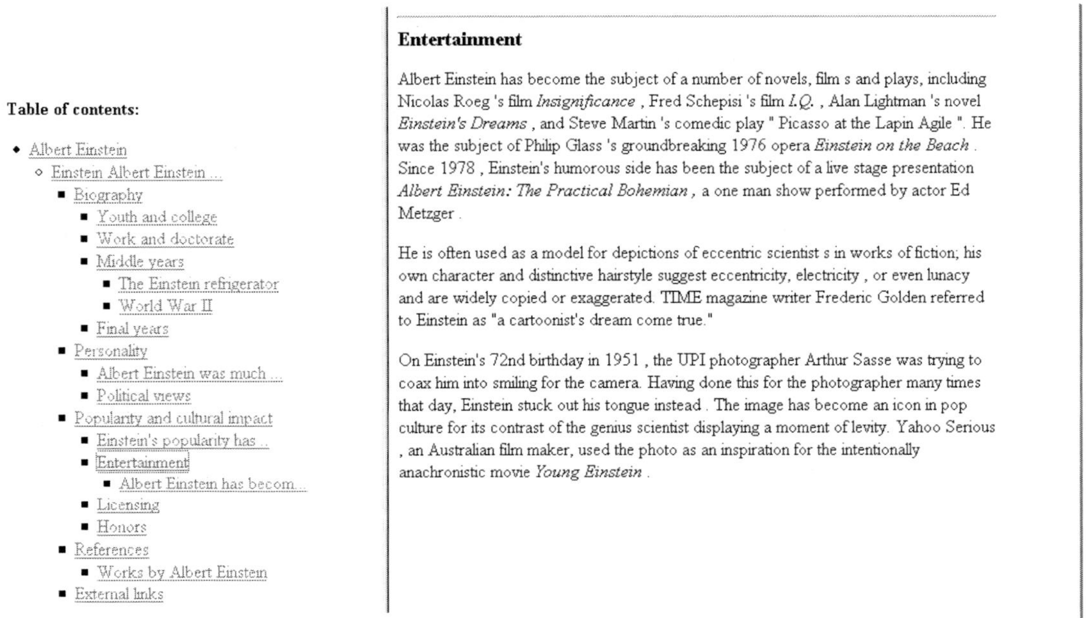

Table of contents:

- Albert Einstein
  - Einstein Albert Einstein ...
    - Biography
      - Youth and college
      - Work and doctorate
      - Middle years
        - The Einstein refrigerator
        - World War II
      - Final years
    - Personality
      - Albert Einstein was much ...
      - Political views
    - Popularity and cultural impact
      - Einstein's popularity has ..
      - Entertainment
        - Albert Einstein has becom...
      - Licensing
      - Honors
    - References
      - Works by Albert Einstein
    - External links

Entertainment

Albert Einstein has become the subject of a number of novels, film s and plays, including Nicolas Roeg 's film *Insignificance* , Fred Schepisi 's film *I.Q.* , Alan Lightman 's novel *Einstein's Dreams* , and Steve Martin 's comedic play " Picasso at the Lapin Agile ". He was the subject of Philip Glass 's groundbreaking 1976 opera *Einstein on the Beach* . Since 1978 , Einstein's humorous side has been the subject of a live stage presentation *Albert Einstein: The Practical Bohemian* , a one man show performed by actor Ed Metzger .

He is often used as a model for depictions of eccentric scientist s in works of fiction; his own character and distinctive hairstyle suggest eccentricity, electricity , or even lunacy and are widely copied or exaggerated. TIME magazine writer Frederic Golden referred to Einstein as "a cartoonist's dream come true."

On Einstein's 72nd birthday in 1951 , the UPI photographer Arthur Sasse was trying to coax him into smiling for the camera. Having done this for the photographer many times that day, Einstein stuck out his tongue instead . The image has become an icon in pop culture for its contrast of the genius scientist displaying a moment of levity. Yahoo Serious , an Australian film maker, used the photo as an inspiration for the intentionally anachronistic movie *Young Einstein* .

Figure 7.2: Presenting an XML element with a table of content (reproduced from [170])

then grouped per document, and the main difference comes from how the documents are ranked. Proposed strategies included calculating directly a document retrieval score [59], using the highest scoring element in the document, or using some sort of aggregated score of the element scores in the document (e.g. average) [199].

A different approach is to start with the ranked list of results produced by the methods described in the previous chapter, which includes overlapping elements [79], and then to group elements per document. Documents are then ranked in the same manner as above described, and then within each document, an overlap removal approach is applied (the brute force filtering is mostly used). Starting from a full list of elements, as opposed to an overlap-free result, is believed to provide a more accurate ranking of the documents since all their relevant contents are taken into account. Note that if documents are ranked according to their highest scoring element, and brute force filtering is applied to remove overlaps, this is the same as starting from an overlap-free result that was generated through brute force filtering.

Most of the approaches described above including variants not described here did, in general, well as long as the initial run upon which they are based (generating the ranked list of elements) performed well. This indicates that properly estimating the relevance of an element is crucial.

Figure 7.3: Presenting XML elements within a document (reproduced from [82])

7.3 ENTRY POINTS

In some applications, it may be useful to provide users with entry points to relevant content in a document, and not a list of relevant elements. For instance, at INEX, an entry point is a location in the document where users should start reading text that is relevant to their information need. A long document may have several entry points for a given query. This was the case for the XML test collection of Shakespeare plays developed by [95]. INEX opted for one entry point per document, referred to as best entry point, since the test collection contained medium-size documents (Wikipedia articles). The best in context task [123] at INEX aimed at identifying relevant documents ranked in order of relevance, and for each relevant document, a pointer to where the most relevant content starts.

Three main strategies were followed to implement the best in context task, depending on the ranked list of results used as input: the ranked list of relevant elements generated by the ranking strategies described in the previous chapter, an overlap-free result list generated by the methods described in Section 7.1, or the result list generated for the relevant in context task at INEX as described in Section 7.2.

For the first two cases, for each document, its element with the highest retrieval score is selected as the best entry point. Documents are ranked on the basis of their selected best entry point retrieval score, e.g. [125, 12]. For the last case (processing a result list produced for the relevant in context task), two strategies were followed. In the first, the document ranking is the one provided by the relevant in context task. In the second strategy, documents are re-ranked. For instance, [199] use the sum of the scores of the elements in a document. This is to place documents with lots of relevant content at the top of the ranking. In both strategies, for each document, the element with the highest retrieval score or its first element is selected as best entry point for that document.

Overall, approaches that identify the best entry point to coincide where relevant content starts to be found in the document did best, as indeed, in the data set used to evaluate the task, it was found that this was the trend [144].

7.4 DISCUSSION

Traditional information retrieval takes as input a document collection and a query, and gives as output a ranked list of documents. No distinction is made between the list of documents estimated relevant by the retrieval model and the documents presented to the users as answers to their queries. This is mainly because there is a fixed unit of retrieval, the document itself. There is, therefore, no reason to question what should be presented to users. In XML retrieval, the general task of estimating the relevance of an element, as accomplished by the ranking strategies described in Chapter 6, also does not question what should be presented to users as answers to their queries. This task, referred to as the thorough task at INEX, is viewed as system-biased, as it only reflects the ability of the XML retrieval system to estimate the relevance of individual XML elements.

XML retrieval goes beyond such a ranked list of results, and this has opened a number of research questions on result presentation, which led to the investigation of various retrieval tasks, each with a particular scenario in mind. The focused task reflects a scenario where a ranked list of XML retrieval is presented to the users with no or little redundant information. The relevant in context task retrieval resembles traditional document retrieval, but where direct access to all relevant content within a document is provided. Finally, the best in context task captures a different notion of relevance, where users desire access to the best information, rather than all relevant information. In this chapter, we described approaches developed at INEX for implementing these retrieval tasks.

Although these tasks are not unrelated, for example, the thorough task is often used as input for further processing for the other tasks, each of these retrieval tasks is capturing a different presentation paradigm in XML retrieval. Each is providing a different view of what approaches are effective for XML retrieval, and how to evaluate their effectiveness. Bringing scenario-specific aspects into information retrieval evaluation has been identified as an important research directions in information retrieval [166].

A final point (an algorithmic one) is the influence of a task to another. For instance, does the approach implementing the focused task affect the outcome of whatever approach is used to implement the relevant in context task? Also, starting for a same list of non-overlapping results, how do the algorithms implementing the relevant in context task compare? Research is needed to answer such questions.

The next chapter describes the evaluation methodology that was adopted by INEX, the INitiative for the Evaluation of XML retrieval. It provides information about the notion of relevance in XML retrieval and its consequence in measuring and comparing XML retrieval systems' effectiveness, including historical background regarding the definition of the above retrieval tasks.

CHAPTER 8

Evaluating XML retrieval effectiveness

It is important to evaluate the benefit of XML retrieval systems to the user. In information retrieval research, the aspect most commonly under investigation is retrieval effectiveness, i.e., the system's ability to satisfy a user's query. For document retrieval systems, this is usually translated to the more specific criterion of a system's ability to retrieve in response to a user query as many relevant documents and as few non-relevant documents as possible.

The predominant approach to evaluate a system retrieval effectiveness is with the use of test collections constructed specifically for that purpose [189, 157, 167]. A test collection usually consists of a set of documents, user requests referred to as topics, and relevance assessments specifying the set of "right answer" for the user requests. There have been several large-scale evaluation projects, which resulted in established information retrieval test collections and evaluation methodology [196, 139, 83]. These, however, are not suitable for the evaluation of XML retrieval as they do not consider the structural information in XML collections. Indeed, they base (mostly) their evaluation on relevance assessments provided at the document level only. In addition, the assumption in IR that documents are independent units (whose relevance is independent of any other document) does not hold in XML retrieval since multiple components retrieved from the same document can hardly be viewed as independent.

To address these and related issues, the INitiative for the Evaluation of XML retrieval (INEX)[1] established in 2002 an infrastructure in the form of large test collections and appropriate evaluation methodology, for evaluating XML retrieval effectiveness. This section provides an overview of the test collections and the evaluation methodology developed in INEX. The document collections, the topics, and the elicitation of the relevance assessments are presented in Sections 8.1 – 8.3, respectively. In addition, all the above are assembled for evaluating particular retrieval tasks. In the context of XML retrieval, a multitude of tasks can be specified, depending on how the relationships between elements are exploited when deciding what to return as answers for given queries (as discussed in Chapter 7). Retrieval tasks are described in Section 8.4. A main research issue in XML retrieval evaluation was the development of appropriate effectiveness measures, which we discuss in Section 8.5.

[1]INEX was funded by DELOS, an EU network of excellence in Digital Libraries. Information about INEX up to 2007 can be found at http://inex.is.informatik.uni-duisburg.de/. Information from 2008 is available at http://www.inex.otago.ac.nz/.

8.1 DOCUMENT COLLECTIONS

The first component of a test collection is the document collection. For XML retrieval evaluation, the collection should consist of documents marked-up in XML. The document collection used in INEX up to 2004 consisted of 12,107 articles marked-up in XML, from 12 magazines and 6 transactions of the IEEE Computer Society's publications, covering the period of 1995-2002, totaling 494 MB in size, and 8 million elements. On average, an article contains 1,532 XML elements, where the average depth of the element is 6.9. In 2005, the collection was extended with further publications from the IEEE Computer Society. Although the collection is relatively small compared with those used in other evaluation initiatives, in particular TREC[2], it has a suitably complex XML structure and the number of elements (the retrieval units) is sufficiently large to carry out realistic experiments for XML retrieval. In 2005, a total of 4,712 new articles from the period of 2002-2004 were added, giving a total of 16,819 articles, totaling 764MB in size and 11 million elements.

Since 2006, INEX has used a different document collection, made from English documents from Wikipedia[3] [40]. The collection consists of the full-texts, marked-up in XML, of 659,388 articles from the Wikipedia project, totaling more than 60 GB (4.6 GB without images) and 52 million elements. The collection has a structure similar to the IEEE collection. On average, an article contains 161.35 XML elements, where the average depth of an element is 6.72. This collection has a richer set of tags (1,241 unique tags compared to 176 in the IEEE collection), and includes a large number of cross-references (represented as XLink).

With Wikipedia, a larger and more realistic test collection was provided. In addition, its content can also appeal to users with backgrounds other than computer science, making the carrying out of interaction studies with this collection more appropriate. However, the documents are rather short (compared to the IEEE documents), which may question whether retrieving document components (i.e. XML elements), instead of whole documents, is a realistic task for this collection. A collection that is being constructed as part of the book search track [90] seems a more appropriate collection for XML retrieval evaluation. Nonetheless, a plentitude of results were obtained using the Wikipedia test collection, advancing state-of-the-art research in XML retrieval.

8.2 TOPICS

To evaluate how effective particular XML retrieval approaches are using a test collection evaluation methodology, statements of information needs are required. These statements are called topics, following TREC terminology [198]. Topics in information retrieval evaluation methodology consist of three main parts [67]: a title field, which consists of a short explanation of the information need, a description field, which consists of a one or two sentence natural language definition of the information need, and a narrative field, which consists of a detailed explanation of the information need and a description of what makes a document (in our case XML elements) relevant or not.

[2] http://trec.nist.org/
[3] http://en.wikipedia.org

From these three fields, titles correspond to instances of queries that would actually be submitted to a retrieval system.

As described in Chapter 4, querying XML documents can be with respect to content and structure. Chapter 4 provided an overview of the development of XML query languages, which had to be reflected in the topics (queries) used for the evaluation of XML retrieval systems. Taking this into account, INEX identified two types of topics:

- Content-only (CO) topics, which are information need statements that ignore the document structure and are, in a sense, similar to the traditional topics used in information retrieval test collections. In INEX, the retrieval results to such topics are elements of various complexity, that is, elements at different levels of the XML documents' structure.

- Content-and-structure (CAS) topics, which are information need statements that refer to both the content and the structure of the elements. These statements might refer to the content of specific elements (e.g., the elements to be returned must contain a section about a particular topic – this is a support condition, see Section 4.1), or might specify the type of the elements to return (e.g., sections should be retrieved – this is a target condition, see Section 4.1).

CO and CAS topics reflect two types of users with varying levels of knowledge about the structure of the searched collection. The first type simulates users who either do not have any knowledge of the document structure or who choose not to use such knowledge. This profile is likely to fit most users searching XML collections. The second type of users aims to make use of any insight about the document structure that they may possess, and then use this knowledge as a precision enhancing device in trying to make the information need more concrete. This user type is more likely to fit expert users, e.g., librarians, patent searchers[4].

As in TREC, an INEX topic consists of the standard title, description, and narrative fields. For CO topics, the title is a sequence of terms. For CAS topics, the title is expressed using NEXI (see Section 4.3.2). Note that NEXI does not allow for the construction of new XML fragments as results (see Section 4.1).

In 2005, the CO topics were extended into Content-Only + Structure (CO+S) topics. The aim was to enable the performance comparison of an XML system across two retrieval scenarios on the same information need, one when structural constraints were expressed (+S) and the other when there were not (CO). The CO+S topics included a CAS title (<castitle>) field, which was a representation of the same information need contained in the <title> field of a CO topic, but including additional knowledge in the form of structural constraints. CAS titles were expressed in the NEXI query language. An example of a CO+S topic is given in Figure 8.1.

An investigation, reported in [184], showed that although improvements in some cases were seen when adding structural constraints to the query, these were not significant. A closer look showed that this was because the structural constraints did not seem to correspond to actual hints; instead

[4]CAS topics can also be created as the outcome of a relevance feedback process, where not only terms but also structural constraints are added to generate the new query (e.g. [161]).

```
<inex_topic topic_id="231" query_type="CO+S">
 <title>markov chains in graph related algorithms</title>
 <castitle>//article//sec[about(.,+"markov chains" +algorithm
                                                        +graphs)]
 </castitle>
 <description>Retrieve information about the use of markov
 chains in graph theory and in graphs-related algorithms.
 </description>
 <narrative>I have just finished my Msc. in mathematics, in
    the field of stochastic processes. My research was in a subject
    related to Markov chains. My aim is to find possible
    implementations of my knowledge in current research. I'm mainly
    interested in applications in graph theory, that is, algorithms
    related to graphs that use the theory of markov chains. I'm
    interested in at least a short specification of the nature of
    implementation (e.g.  what is the exact theory used, and to
    which purpose), hence the relevant elements should be sections,
    paragraphs or even abstracts of documents, but in any case,
    should be part of the content of the document (as opposed to,
    say, vt, or bib).
 </narrative>
</inex_topic>
```

Figure 8.1: A CO+S topic from the INEX 2005 test collection

they appeared to be a function of the document collection (i.e. the structural constraints tended to be based on the most frequent and meaningful tags in the collection) rather than the query. At this stage, it is not clear if this is true of only the given collection or all collections. Further investigation is ongoing.

Regarding CAS topics (including their +S counterparts), the actual definition of CAS topics has not changed over the years. CAS topics are topic statements that contain explicit references to the XML structure. More precisely, a CAS query contains two kinds of structural constraints: where to look (i.e. the support elements), and what to return (i.e. the target elements). What has evolved over the years is how to interpret the structural constraints, since each structural constraint could be considered as a strict (must be matched exactly) or vague (do not need to be matched exactly) criterion. The former interpretation closely reflects the database (data-oriented) view, where only records that exactly match the specified structure should be retrieved. The latter is closer to the information retrieval (content-oriented) view, where users' information need is assumed to be inherently uncertain or vague. (Please refer to Section 3.5 for a discussion on these two views).

An investigation reported in [185] looked at whether the two interpretations of the structural constraints mattered when evaluating the performance of XML retrieval systems. In a CAS topic, depending on how the target elements and/or the support conditions are treated, four interpretations are possible: when the structural constraints of both target and support elements are vague,

year	number of selected topics	number of assessed topics
2002	60	54
2003	66	62
2004	71	60
2005	87	63
2006	125	114
2007	130	99
2008	135	70
Total	674	522

when the structural constraints of the target element is vague and that of the support elements is strict, etc. Results suggest that, in terms of comparing retrieval effectiveness, there are two separate interpretations of CAS that matter, one in which the target element is interpreted strictly and the other in which it is interpreted vaguely. The interpretation of the support conditions does not appear to be important. These results have implication in how to assess the relevance of elements, as there is no need to worry whether the structural constraint in the support condition should be assessed strictly or vaguely, which simplifies the assessment process greatly.

In INEX, the topics are created from the participating groups. Each year, each participating group is asked to submit up to a given number of candidate topics, following a detailed guideline regarding the creation of meaningful topics for evaluating XML retrieval effectiveness. INEX then selects the topics to form the collection (each year). The selection is based on a combination of criteria such as balancing the number of topics across all participants (as the aim is for participants to assess their own topics - see Section 8.3), eliminating topics that are considered too ambiguous or too difficult to judge, and uniqueness of topics. Table 8.2 gives statistics about the topics that were collected in each round of INEX[5]. We also show the number of topics assessed to date. The yearly INEX proceedings contain more statistics on the topics [47, 53, 55, 54], and [80] provides some analysis of the type of structural constraints found in the CAS topics.

8.3 RELEVANCE ASSESSMENTS

An information retrieval system is effective if it retrieves all and only the relevant documents for a given query. To measure this, we require for each topic the list of relevant documents, referred to as relevance assessments. In the context of XML retrieval, relevance assessments consist of the set of relevant elements for each topic. We discuss relevance in XML retrieval, and then the methodology adopted by INEX to gather relevance assessments.

[5] In 2008, additional 125 topics were selected from a proxy log. Their assessments were based on the wikipedia pages clicked for these topics [60].

Most dictionaries define relevance as "pertinence to the matter at hand". In information retrieval, it is usually understood as the connection between a retrieved item and the user's query. In XML retrieval, the relationship between a retrieved item and the user's query is further complicated by the need to consider the structure in the documents. Since retrieved elements can be at any level of granularity, an element and one of its child elements can both be relevant to a given query, but the child element may be more focussed on the topic of the query than its parent element, which may contain additional irrelevant content. In this case, the child element is a better element to retrieve than its parent element, because not only it is relevant to the query, but it is also specific to the query. To accommodate the specificity aspect, INEX defined in 2002 relevance along two dimensions:

- Topical relevance, which reflects the extent to which the information contained in an element satisfies the information need, i.e. measures the exhaustivity within an element about the topic.
- Component coverage, which reflects the extent to which an element is focussed on the information need, and not on other, irrelevant topics, i.e. measures the specificity of an element with regard to the topic.

A multiple degree relevance scale was necessary to allow the explicit representation of how exhaustively a topic is discussed within an element with respect to its child elements. For example, a section containing two paragraphs may be regarded more relevant than either of its paragraphs by themselves. Binary values of relevance cannot reflect this difference. INEX, therefore, adopted a four-point relevance scale based on [101]:

- Irrelevant (0): The element does not contain any information about the topic of request.
- Marginally relevant (1): The element mentions the topic of request, but only in passing.
- Fairly relevant (2): The element discusses the topic of request, but not exhaustively.
- Highly relevant (3): The element discusses the topic of request exhaustively.

As for topical relevance, a multiple scale was also necessary for the component coverage dimension. This is to allow to reward retrieval systems that are able to retrieve the appropriate ("exact") sized elements. For example, a retrieval system that is able to locate the only relevant section in a book is more effective than one that returns a whole chapter. A four-point relevance scale for component coverage was, therefore, also adopted:

- No coverage (N): The topic, or an aspect of the topic, is not a theme of the element.
- Too large (L): The topic, or an aspect of the topic, is only a minor theme of the element.
- Too small (S): The topic, or an aspect of the topic, is the main or only theme of the element, but the component is too small to act as a meaningful unit of information.
- Exact coverage (E): The topic, or an aspect of the topic, is the main or only theme of the element, and the element acts as a meaningful unit of information.

Based on the combination of the topical relevance and component coverage, it becomes possible to identify those relevant elements, which are both exhaustive and specific to the topic of request and hence represent the most appropriate units to return to the user. With this definition of relevance, it becomes possible to reward systems that are able to retrieve these elements.

The component coverage dimension allows the classification of components as "too small" if they do not bear self-explaining information for the user and thus cannot serve as informative units. However, in a study of the collected assessments for 2002, the use of "too small" led to some misinterpretations while assessing the coverage of an element [98]. The problem was that, for CAS topics, the "too small" and "too large" coverage categories were incorrectly interpreted as the relation between the actual size of the result element and the size of the target element instead of the relation between the relevant and irrelevant contents of the result element. To address this, INEX 2003 renamed the two dimensions to exhaustivity and specificity, based on Chiaramella et al. work on structured document retrieval [27]:

- Exhaustivity, which measures how exhaustively an element discusses the topic of the user's request.
- Specificity, which measures the extent to which an element focuses on the topic of request (and not on other, irrelevant topics).

The scale for the exhaustivity dimension, which replaces the topical relevance dimension, was redefined by simply replacing the word relevant to exhaustive. To avoid direct association with element size, the specificity dimension, which replaces the component coverage, adopted an ordinal scale similar to that defined for the exhaustivity dimension:

- Not specific (0): the topic of request is not a theme discussed in the element.
- Marginally specific (1): the topic of request is a minor theme discussed in the element.
- Fairly specific (2): the topic of request is a major theme discussed in the element.
- Highly specific (3): the topic of request is the only theme discussed in the element.

Throughout the yearly INEX campaigns, there have been arguments against the separation of relevance into two dimensions. It was, however, believed that this separation was needed in order to provide a more stable measure of relevance than if assessors were asked to rate elements on a single scale. One reason for this is that assessors are likely to place varying emphasis on these two dimensions when assigning a single relevance value. For example, one assessor might tend to rate highly specific elements as more relevant, while another might be more tolerant of lower specificity and prefer high exhaustivity.

However, obtaining relevance assessments is a very tedious and costly task [143]. An observation made by Clarke [30] was that the assessment process could be simplified if first, relevant passages of text were identified by highlighting, and then the elements within these passages were assessed. As a consequence, at INEX 2005, the assessment method was changed, leading to the redefinition of the scales for specificity. The procedure was a two-phase process. In the first phase, assessors highlighted text fragments containing only relevant information. The specificity dimension was then automatically measured on a continuous scale [0,1], by calculating the ratio of the relevant content of an XML element: a completely highlighted element had a specificity value of 1, whereas a non-highlighted element had a specificity value of 0. For all other elements, the specificity value was defined as the ratio (in characters) of the highlighted text (i.e. relevant information) to the element size. For example, an element with specificity of 0.72 has 72% of its content highlighted.

In the second phase, for all elements within highlighted passages (and parent elements of those), assessors were asked to assess their exhaustivity. Following the outcomes of extensive statistical analysis performed on the INEX 2004 results [134], which showed that in terms of comparing retrieval effectiveness, the same conclusions could be drawn using a smaller number of grades for the exhaustivity dimension[6], INEX 2005 adopted the following 3 + 1 exhaustivity values:

- Highly exhaustive (2): the element discussed most or all aspects of the query.
- Partly exhaustive (1): the element discussed only few aspects of the query.
- Not exhaustive (0): the element did not discuss the query.
- Too Small (?): the element contains relevant material but is too small to be relevant on it own.

The category of "too small" was introduced to allow assessors to label elements, which although contained relevant information, were too small to sensibly reason about their level of exhaustivity. In 2002, the "too small" category was with respect to the specificity aspect of relevance, whereas in 2005, it is a degree of exhaustivity, and was deemed necessary to free assessors from the burden of having to assess very small text fragments whose level of exhaustivity could not be sensibly decided.

A continuous discussion in INEX was whether the exhaustivity dimension was needed. Using one dimension would be less costly to obtain. In addition, assessors felt that gauging exhaustivity was a cognitively difficult task to perform, and the extra burden led to less consistent assessments [178] compared to those obtained at TREC. An extensive statistical analysis was, therefore, performed on the INEX 2005 results [134], which showed that in terms of comparing retrieval performance, not using the exhaustivity dimension led to similar results in terms of comparing retrieval effectiveness. As a result, INEX 2006 dropped the exhaustivity dimension[7]. Relevance was defined only along the specificity dimension.

Each year, assessors, which were the INEX participants, provided the relevance assessments through the use of an online relevance assessment tool, called X-RAI [143][8]. The purpose of the interface was to both ensure sound and complete relevance assessments, and to ease the assessment process. This was crucial because the assessors are the INEX participants, who are not paid to perform the task, but do it in addition to their ordinary duties.

As in TREC, a pooling method [68] is used to elicit the elements to assess. Participants are asked, for each topic, to submit a ranked list of up to 1,500 elements, in order of estimated relevance. At INEX 2002 the top 100 elements from each ranked list were taken from each run to form the

[6]The same observation was reached for the specificity dimension, but as the assessment procedure was changed in INEX 2005, the new highlighting process allowed for a continuous scale of specificity to be calculated automatically, and hence to be more reliable.

[7]To be correct, the grades of exhaustivity became binary; a highlighted element was exhaustive (relevant), whereas a non-highlighted element was not exhaustive (not relevant).

[8]In 2002, a rather cumbersome on-line tool was used, which was replaced in 2003 by X-RAI.

pool. The assessor was then asked to assess the pool on a document-by-document basis. A criticism of this method was that the pool for a given topic may be covered by a single document. As assessors are assessing on a document by document basis, it was decided in 2003 that each pool should contain the same number of documents and not elements [143]. For this purpose, the top-n elements (for increasing values of n) are taken from each submitted run until there are over 500 documents in the pool. This method has been used since 2003, and seems satisfactory (but see [144] for a detailed investigation about the suitability of this pooling method).

For 2003 and 2004, the online assessment tool displayed the document with elements from the runs identified with a dotted box (see Figure 8.2). Assessors were asked to click on each box and assign a relevance value to the boxes. A pop-up window displayed all the possible relevance values allowed for an element (a maximum of ten[9]). For each element the currently assigned relevance values were displayed in front of the XML tag name.

For the CAS topics, the relevance of an element was to be assessed with respect to the content criterion only. This was because structural conditions were viewed as hints into where an XML retrieval should look to find relevant information. In addition, the obtained assessments could be used for evaluating retrieval tasks (see Section 8.4) where the structural constraints are to be strictly interpreted. As discussed in Section 8.2 regarding the interpretation of the structural constraints [185], what matters is how to interpret the constraints (i.e. strict vs. vague) with respect to the target elements (what elements to return). Therefore, the set of assessments can be filtered to derive the relevance assessments that strictly match the target elements[10].

In the assessment tool, enforcement rules were implemented to decide which element was to be assessed (e.g. if a given element in the pool was assessed relevant then its parents and descendants were added to the pool, or if an element was assessed irrelevant then its descendants were removed from the pool), and what possible relevance values an element could take (e.g. if an element was assessed as relevant then it could not be more exhaustive than its parent). By INEX 2004, the enforcement rules had became too demanding and obstructive for the assessors, who were by then obliged to explicitly assess most of the elements of a document if it contained any relevant elements [115].

This, however, changed in 2005 with the new assessment procedure. Unlike previous INEX campaigns, the elements identified by the XML retrieval systems were shown, but assessors were asked to identify relevant passages rather than to make decisions as to the relevance of these identified elements. In the first pass, assessors highlight text fragments (or passages) that contain only relevant information. Highlighting was based solely on relevance irrespective of the two dimensions and their scales. Assessors were asked not to highlight larger contexts if they contained irrelevant fragments. In the second phase, assessors were asked to assess only the exhaustivity of relevant elements, i.e. those elements that intersect with any of the highlighted passages. In the interface, the highlighted

[9]For example, an element could not be assessment as not specific and partially exhaustive at the same time.
[10]This assessment strategy was adopted in 2003 for simplicity, and turned out to be valid, as discussed in Section 8.2 [185].

Figure 8.2: INEX 2004 assessment interface

passages were shown in yellow, and for each assessed element, the currently assigned exhaustivity value was displayed in front of the XML tag name.

This new procedure reduced considerably the number of enforcement rules, resulting in a much more natural and non intrusive way of assessing. From 2006, the assessors had only to highlight relevant text fragments from articles identified as candidates using the pooling strategy. Figure 8.3 shows a screen shot of the 2007 interface. This change also has positive consequences on two important factors of assessments: the completeness (i.e. the proportion of relevant elements that have been assessed) and the soundness (i.e. the reliability of the assessments) [144]. Now, after several years of experimentation, it is believed that the highlighting procedure to gather assessments produces sound and sufficiently complete assessments for XML retrieval evaluation.

8.4 RETRIEVAL TASKS

Retrieval systems use indexing and scoring mechanisms to select what results to return as answers to queries, and in what order. These results are then evaluated according to how relevant they are to the given queries. In the test collection evaluation methodology, this is simulated through the specification of so-called retrieval tasks to be performed by the retrieval systems. It is the outcomes of these retrieval tasks that get evaluated.

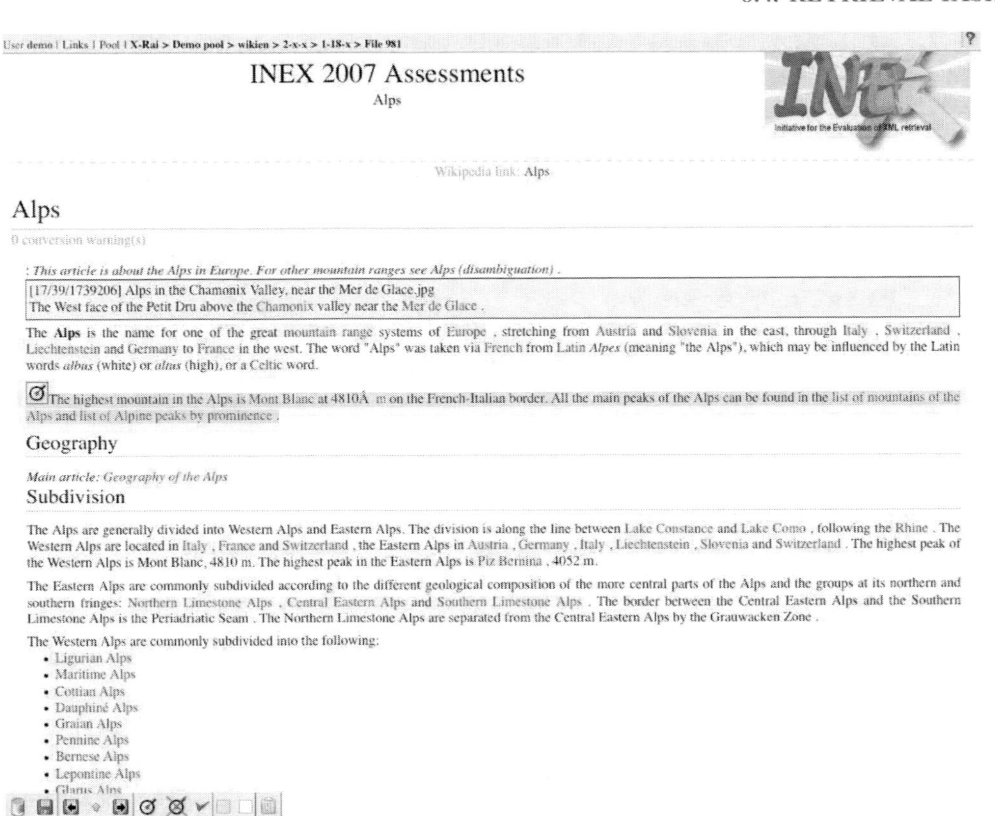

Figure 8.3: INEX 2007 assessment interface

The typical retrieval task investigated in information retrieval is the ad hoc task. In information retrieval literature [196], ad hoc retrieval is described as a simulation of how a library might be used and involves the searching of a static set of documents using a new set of topics. In traditional information retrieval, the library consists of documents, representing well-defined units of retrieval. The user's information need is typically expressed in the form of a set of terms. Given this, the task of an information retrieval system is to return to the user, in response to his/her query, as many relevant documents and as few irrelevant documents as possible. The output is usually presented to the user as a ranked list of documents, ordered by presumed relevance to the query.

Ad hoc retrieval in XML retrieval can be described in a similar manner to traditional information retrieval: a static collection being searched using new topics. The difference is that, here, the collection consists of XML documents, composed of different granularity of nested XML elements, each of which representing a possible unit of retrieval. The user's query may also contain structural conditions. As presented in Chapter 7, the output of an XML retrieval system may follow the traditional ranked list presentation, or may extend to non-linear forms, such as grouping of elements per document.

Up to 2004, ad-hoc retrieval was defined as the general task of returning, instead of whole documents, those XML elements that are most specific and exhaustive to the user's query. In other words, systems should return components that contain as much relevant information and as little irrelevant information as possible. From 2006, the right level of granularity was defined upon specifity only. Within this general task, several sub-tasks were defined, where the main difference was the treatment of the structural constraints.

The CO sub-task makes use of the CO topics, where an effective system is one that retrieves the most specific elements and only those, which are relevant to the topic of request. Here, there are no structural constraints. The CAS sub-task makes use of CAS topics, where an effective system is one that retrieves the most specific document components, which are relevant to the topic of request and match, either strictly or vaguely, the structural constraints specified in the query. In 2002, a strict interpretation of the CAS structural constraints was adopted, whereas in 2003, both, a strict and a vague interpretations was followed, leading to the SCAS sub-task (strict content-and-structure), defined as for the INEX 2002 CAS sub-task, and the VCAS sub-task (vague content-and-structure). In that last sub-task, the goal of an XML retrieval system was to return relevant elements that may not exactly conform to the structural conditions expressed within the query, but where the path specifications should be considered hints as to where to look. In 2004, the two sub-tasks investigated were the CO sub-task, and the VCAS sub-task. The SCAS sub-task was felt to be an unrealistic task because specifying an information need is not an easy task, in particular for XML collections possessing a wide variety of tag names.

However, within this general task, the actual relationship between retrieved elements was not considered, and many systems returned overlapping elements (e.g. nested elements). Indeed, the top 10 ranked systems for the CO sub-task in INEX 2004 contained between 70% to 80% overlapping elements. What most systems did was to estimate the relevance of XML elements (using the strategies described in Chapter 6), which is different to identifying the most relevant (i.e. most exhaustive and specific) elements. This had very strong implications with respect to measuring effectiveness, where approaches that attempted to implement a more selective approach (e.g., between two nested relevant elements, returning the one most specific to the query) performed poorly. In addition, evidence from the investigation of user behaviour in XML retrieval, supported the view that users did not like being returned (at least too many) overlapping elements [176].

As a result, the focussed sub-task was defined in 2005, intended for approaches concerned with the so-called focused retrieval of XML elements, i.e. aiming at indeed targeting the appropriate level of granularity of relevant content that should be returned to the user for a given topic. The aim was for systems to find the most exhaustive and specific element on a path within a given document containing relevant information and return to the user only this most appropriate unit of retrieval. In 2006, this was translated to returning the most specific element on a path within a given document.

Returning overlapping elements was not permitted. Section 7.1 describes approaches developed to implement this task.

The INEX ad-hoc general task, as carried out by most systems up to 2004, was renamed in 2005 as the thorough sub-task. At the beginning of INEX 2006, the usefulness of the thorough sub-task was questioned. Nonetheless, this is an important task, because most if not all of the other sub-tasks require an estimation of the relevance of all elements, even if not all of them will actually be presented to the user; the other sub-tasks are indeed post-processing (in one way or the other) the ranked list of results generated by the thorough sub-task, each leading to a different way to present results in XML retrieval (see Chapter 7)[11].

Within all the above sub-tasks, the output of XML retrieval systems was assumed to be a ranked list of XML elements, ordered by their presumed relevance to the query, whether overlapping elements were allowed or not. However, user studies [177] suggested that users were expecting to be returned elements grouped per document, and to have access to the overall context of an element. The fetch & browse task was introduced in 2005 for this reason. The aim was to first identify relevant documents (the fetching phase), and then to identify the most exhaustive and specific elements within the fetched documents (the browsing phase). In the fetching phase, documents had to be ranked according to how exhaustive and specific they were. In the browsing phase, ranking had to be done according to how exhaustive and specific the relevant elements in the document were, compared to other elements in the same document. In 2005, no explicit constraints were given regarding whether returning overlapping elements within a document was allowed. In 2006, the same task, renamed the relevant in context sub-task, required systems to return for each document an unranked set of non-overlapping elements, covering the relevant material in the document. The idea here was to respect the reading order within a document (i.e. not ranking of elements within a document). Section 7.2 describes approaches developed to implement the relevant in context sub-task.

In addition, a new task was introduced in 2006, the best in context sub-task, where the aim was to find the best-entry-point, here a single element, for starting to read documents with relevant information. This sub-task can be viewed as the extreme case of the fetch & browse approach, where only one element is returned per document. Section 7.3 describes approaches developed to implement this task.

A continuous question at INEX was whether all the above tasks were realistic. In other words, the question is who are the users for which these tasks would make sense [178]. The focused sub-task, classified as a user-oriented task, was, for instance, criticized for being context-free [135] – that is, the user is presented with focused elements, but cannot evaluate the authority of the information because no context is given. As discussed in Section 7.2, a table of content can be shown in addition to the focused element to provide some context.

A use case studies track was set up in INEX 2006 to investigate these issues [182]. It was found that several commercial systems implemented the tasks investigated at INEX, although whether

[11]The effectiveness of this sub-task was, however, not measured since 2007, although its inclusion is currently being discussed for INEX 2009.

XML was used as the document mark-up is not clear [182, 44]. One type of applications, book search (e.g. Books27x7, Safari, and Google Book Search), was found to be implemented in various ways, including as a (surprisingly) thorough retrieval sub-task, and a fetch & browse sub-task. There is currently a book search track at INEX [90].

Since 2007, INEX has a passage retrieval task where the aim is to identify the appropriate size of results to return and their location. Passage retrieval has been studied in information retrieval around the mid 90s, e.g., [24, 70, 86, 158, 201] (see also Section 3.2). The idea for the passage retrieval task came from the way the assessment process has been carried out since 2005, i.e., highlighting the relevant content in a document. Thus passage retrieval is nothing else than attempting to identify these highlighted fragments, and to rank them appropriately. An interesting question is whether identifying these best passages first, and then translating them into elements is a better strategy than identifying directly the best elements to return as answer to queries [73, 75]. Another question, currently being investigated, is comparing the effectiveness of passage retrieval to the above sub-tasks [60].

8.5 MEASURES

Given a document collection, a set of topics, and relevance assessments, measures are used to evaluate how effective is an information retrieval system at retrieving relevant information for a given retrieval task. In this section, we only discuss the measures used to evaluate the thorough and focused retrieval sub-tasks. Measures evaluating the other tasks can be found in [108, 81].

Since its launch in 2002, INEX has been challenged by the issue of how to measure an XML retrieval system's effectiveness. The main complication comes from the necessity to consider the dependency between elements when evaluating effectiveness. Unlike traditional information retrieval, users in XML retrieval have access to other, structurally related elements from returned result elements. They may hence locate additional relevant information by, for example, browsing or scrolling. Obviously, different user interfaces and results presentations will support different forms of user interaction, where the ease of access to structurally related components will vary. Nevertheless, it is important to consider that relevant information can be accessed through structurally related elements when evaluating XML retrieval effectiveness.

This motivates the need to consider so-called near-misses, which are elements from where users can access relevant content, within the evaluation. The alternative, to ignore near-misses, would lead to a much too strict evaluation scenario. For example, assuming that paragraph `p` is the most relevant element of document `d` for a given query, returning the section `s` containing that paragraph corresponds to a near-miss, since the user will be able to access `p` by "browsing down" from `s`. We also have the other way round, the most relevant element being `s`, and the paragraph `p` being returned, as the user will be able to access `s` by "browsing up" from `p`. Similar scenarios were also considered at the web track of TREC-8 [69], e.g. by taking into account web pages linking to relevant content when evaluating retrieval effectiveness.

Ad hoc retrieval effectiveness is usually evaluated by the established and widely used precision and recall measures and their variants (Chapter 3, [197]). From 2002 to 2004, INEX used inex_eval [65], which applies the measure of precall [148] to XML elements. As for traditional precision and recall measures, inex_eval is based on a counting mechanism, i.e., on the number[12] of retrieved and relevant elements. As a consequence, if we consider near-misses when evaluating retrieval effectiveness, systems that return relevant but overlapping elements will be evaluated more effective than those that do not return overlapping elements. If, for instance, both a relevant paragraph and its enclosing section are retrieved, then both elements will be counted as separate relevant components (the fact that the paragraph is relevant means that the section is also relevant although likely to a different extent), which increases the count of relevant and retrieved elements. Therefore, despite not retrieving entirely new relevant information, systems that retrieve overlapping relevant elements would receive higher effectiveness scores [94].

The first step to address this problem - as presented in Section 8.4 - was to define the two retrieval tasks, thorough and focussed, to distinguish between systems that were interested in estimating the relevance of elements given a topic of request, and those that aimed at providing so-called focused access to XML content. Using the inex_eval measure to evaluate the thorough sub-task is then appropriate, although it is preferable for both simplicity and consistency to use the same measure or family of measures to evaluate all tasks. As a consequence, a plethora of evaluation measures has been proposed for XML retrieval, e.g., T2I [37], inex_eval_ng [62], PRUM [142], XCG [93], and HiXEVAL [138]. We only briefly describe the latter two.

INEX adopted in 2005 a new set of measures, called XCG, for both sub-tasks [93]. The XCG measures are an eXtension of the Cumulated Gain based measures [76]. These measures are not based on a counting mechanisms, but on cumulated gains associated with returned results. These were developed specifically for graded (non-binary) relevance and with the aim to allow information retrieval systems to be compared according to the retrieved documents gain values. The retrieval effectiveness of a system is calculated as a comparison of the gain values obtained by the system and that of an ideal system. The ideal system is obtained by ranking all the documents (from the relevance assessments) in order of their gain values. For XML retrieval, the gain is based on the combination of the two relevance dimensions used at INEX, where the highly (3) exhaustive and specific results give the highest gain. The combination was expressed through so-called quantization functions [92].

With respect to the focussed task, as we still want to appropriately reward the retrieval of near-misses, we need to differentiate between those elements that should be retrieved, i.e. the desired elements, and those elements that are structurally related to the desired elements, i.e. the near-misses. It was, therefore, necessary to separate between these two sets by marking a subset of the relevant elements in the recall-base as ideal answers, i.e. the so-called desired elements. We refer to this set as the ideal recall-base. It should be pointed out that using inex_eval on the ideal-recall base to evaluate

[12] With non-binary relevance, it is the relevance grade of each retrieved element that is added in the count.

the focussed task would mean that near-misses would not be considered when evaluating retrieval performance.

The short-coming of XCG, as defined at INEX, is the construction of the ideal-recall base. A study reported in [88] shows that the chosen methodology can impact on the obtained performance scores, which is not an easy problem to solve. This and also with the way the assessments were gathered (the highlighting procedure) led to using a different family of measures. These are an extension of the traditional definitions of precision and recall that include the knowledge obtained from the highlighting assessment procedure adopted at INEX 2005. These measures were initially proposed by Pehcevski and Thom [138], i.e. HiXEVAL, finalized in [81], and have since 2007 been used as the official INEX retrieval effectiveness measures.

We recall that the aim of an XML retrieval system is to return those elements that contain as much relevant information as possible, and as little non-relevant information as possible. With the highlighting assessment procedure, this translates to the aim of returning elements that contain as much highlighted (relevant) content as possible, and as little non-highlighted (non-relevant) content as possible. The classical definition of precision and recall can be modified to reflect this view:

$$Precision = \frac{amount\ of\ relevant\ information\ retrieved}{total\ amount\ of\ information\ retrieved}$$

$$Recall = \frac{amount\ of\ relevant\ information\ retrieved}{total\ amount\ of\ relevant\ information}$$

Instead of counting the number of relevant items (in our case elements) retrieved, we are measuring the amount of relevant (highlighted) text retrieved, which solves the problem of the impact of two nested elements, i.e. both being counted, in the evaluation.

More formally, let e_i be an element retrieved at rank i. Let $rsize(e_i)$ be the amount of highlighted (relevant) text contained in e_i for a given topic (if there is no highlighted text in the element, $rsize(e_i) = 0$). Let $size(e_i)$ be the total number of characters contained in e_i, and let $Trel$ be the total amount of (highlighted) relevant information in the collection for the topic. Precision at rank r corresponds to the fraction of retrieved relevant information up to rank r:

$$P@r = \frac{\sum_{i=1}^{r} rsize(e_i)}{\sum_{i=1}^{r} size(e_i)} \tag{8.1}$$

This definition ensures that, to achieve a high precision value at rank r, the elements retrieved up to and including that rank need to contain as little non-relevant information as possible (i.e. maximizing $rsize(e_i)$ for each e_i).

Recall at rank r corresponds to the fraction of relevant information retrieved up to rank r:

$$R@r = \frac{1}{Trel} \cdot \sum_{i=1}^{r} rsize(e_i) \tag{8.2}$$

This definition ensures that, to achieve a high recall value at rank r, the elements retrieved up to and including that rank need to contain as much relevant information as possible.

The definition of $Trel$ depends on the task. For the thorough case, $Trel$ is the total number of highlighted characters across all elements. For the focused case, $Trel$ is the total number of highlighted characters across all documents. The difference between the two cases is that, in the former case, overlapping text (contained by all the overlapping relevant elements) is used to calculate the total amount of highlighted relevant information for the topic, whereas for the latter case, non-overlapping text is used.

An issue with the precision measure $P@r$ is that it can be biased towards systems that return several shorter document parts rather than returning one longer part that contains them all. Therefore, INEX instead uses precision at recall levels rather than at ranks. Specifically, INEX uses an interpolated precision measure $iP[x]$, which calculates interpolated precision scores at selected recall levels x. This allows measuring which system is more capable of retrieving as much relevant text as possible (the selected recall level), without also retrieving a substantial amount of non-relevant text.

As for traditional precision and recall in information retrieval evaluation, mean average precision has been defined (see [81] for the full details).

Finally, as INEX is incorporating other types of focused retrieval tasks, these measures, which are defined on the amount of highlighted text, seem appropriate to evaluate these other tasks, in particular the new passage retrieval task introduced at INEX 2007 [52].

8.6 DISCUSSION

At the time of writing, the INEX campaign has been running for seven consecutive years [47, 52, 53, 54, 55, 56, 52, 60], and is now in its eight year. Evaluating XML retrieval was a challenge as the addition of the structure led to many complex issues, which were not always identified at the beginning (e.g., the issue of overlap and a counting-based measure, the difficulty in consistently assessing elements using a four-graded two-dimension scale). In addition, limited funding was available, for example, to pay assessors. This led to a research problem in itself, e.g., how to elicit quality assessments in order for the test collections to be reusable [144]. After six years, and a number of changes (e.g., the definition of relevance in XML retrieval, a plethora of measures), some stability was achieved in 2007, which was welcomed by the XML retrieval community. It also means that we can start investigating other research questions in XML retrieval and evaluation.

To date, most XML retrieval approaches developed during the INEX campaigns have been concerned with the retrieval of flat text units such as elements and recently passages. An important omission has been the retrieval of trees from (hierarchically) structured text. For example, XML query languages such as XQuery Full-Text include support for queries with tree-based structural constraints and specifications for tree-based results. We recall that INEX used NEXI as its query language, thus with no possibility to construct result trees. A number of approaches, for example, XRANK [66], XXL [172], and XIRQL [49], output ranked lists of XML document subtrees.

Their full capability cannot yet be evaluated. With the growing interest in tree retrieval, evaluation methodologies are required to measure tree retrieval effectiveness. Some initial work can be found in [2].

In the relevant in context task, elements from the same documents are grouped together. There is nothing stopping the creation of virtual documents, i.e., new documents made from some intelligent aggregation of elements coming from different documents, for example using XQuery Full-Text. Here, we are not limited to XML retrieval; other focused retrieval tasks may be involved (see Section 3.9). Indeed, passages and answers as returned by passage retrieval systems and query answering systems, respectively, may also be used to form a virtual document. Even full documents may be used, as already proposed by major search engines with their so-called aggregated page result, e.g., Alpha Yahoo! or Google Universal search. Again the task of creating virtual but meaningful documents will require new evaluation methodologies. The work of [2] may also be of help here.

Although not discussed here, as for TREC, INEX runs several additional tracks studying different aspects and tasks of XML retrieval, such as interaction, relevance feedback, searching heterogeneous collections, entity ranking, natural language query, question & answering, and efficiency, to name a few. We briefly introduce them in the next chapter, which concludes this book.

CHAPTER 9

Conclusions

A main difference between XML retrieval and classic "flat" information retrieval is that in the former, the aim is to retrieve the most relevant elements, whereas the latter is concerned with retrieving the most relevant documents. These two aims can be related, as witnessed with one of the retrieval tasks investigated at INEX, namely, the relevant in context task.

XML retrieval exploits the XML mark-up to identify and return the most relevant and specific document components as answers to a query, i.e. XML elements at the right level of granularity. This book described query languages, indexing strategies, ranking algorithms and presentation paradigms developed for XML retrieval. The book also described the evaluation methodology developed by INEX to measure XML retrieval effectiveness. We reflect upon these in Section 9.1.

The goal of returning elements at the right level of granularity as answers to a query has recently been coined as XML "element" retrieval [182]. XML retrieval has other research challenges that go beyond identifying what are the most relevant elements to a given user query. Some of the research challenges have been the focus of dedicated tracks at INEX, which we briefly introduce in Section 9.2. This chapter ends in Section 9.3 by positioning XML retrieval in the bigger picture of information access.

9.1 XML "ELEMENT" RETRIEVAL

One way to exploit the XML mark-up is by writing queries that pose conditions on both the content and the structure of the elements to be retrieved. XQuery Full-Text is a query language for XML retrieval, which is currently developed by the W3C. XQuery Full-Text aims to address both database and information retrieval information needs. Coming from the information retrieval community is NEXI, which is an enhanced subset of XPath. The enhancement comes with the addition of the aboutness function, deemed more appropriate for content-oriented information needs. NEXI has been developed so that researchers worldwide participating at INEX had a common language for expressing the structural aspect of the information need. NEXI has served this purpose well. An interesting question is the influence of NEXI on any further development of XQuery Full-Text, and the adoption and deployment of these two languages in industry, for instance, in the patent domain and the publishing world.

One unsolved issue is the usefulness of content-and-structure queries. Indeed, using content-and-structure queries in INEX did not lead, contrary to early expectations, to improved retrieval performance [185]. We return to this later in this section.

What is the most effective indexing strategy for XML retrieval is not clear. Apart for the selective strategy, as long as the term statistics properly reflect how good a term is at indexing an element, compared to another term, the effect of a particular indexing strategy (and associated ranking strategy) is likely to be minimal. The choice of an indexing strategy, whether element-based, leaf-only, aggregation-based, or distributed, is an efficiency problem, and not an effectiveness one. This is not the case for the selective indexing strategy, which can have an effect on effectiveness, e.g. [164]. This is because this strategy is concerned with deciding what constitute potential units of retrieval (e.g. elements above a certain size, elements of selected types). As a matter of fact, the selective indexing strategy has often been used with the other indexing strategies for this exact purpose.

The issue of nested elements and term statistics (mostly inverse element frequency) led to several ways to estimate the inverse element frequency of a term. This issue seems less important to what initially thought, if it is at all an issue. What seems to matter is the relative inverse element frequency of a term compared to another, and not on which basis (all elements, elements of a given type, etc) the inverse element frequency of a term is calculated.

Estimating an element relevance can be viewed as a combination of evidence problem, which has two sides: what combination framework to use and which evidence to use in the combination. In information retrieval, uncertainty theories [104] have often been used as the combination framework. There are three main trends, namely, probability theory, Dempster-Shafer theory of evidence and fuzzy set theory. All have been applied to XML retrieval (see e.g. [153, 106, 109, 21, 132, 96, 97, 27, 154, 48, 165, 117], although these have not all been experimented with during the INEX campaigns). As is the case in many other areas of information retrieval, frameworks (retrieval models) based on probability theory have often been used in XML retrieval. A recent trend in information retrieval is the use of machine learning as the combination framework. The aim here is to train a parametric function, for example, using assessment data, to derive an appropriate scoring function that will then be used to estimate an element relevance. An example for XML retrieval is [192]. A more recent trend, particularly pertinent to XML retrieval, is to learn the scoring function among a set of possible scoring functions, referred as learning to rank [77].

With respect to choosing which evidence to combine, considering the element content, its size, and its context has shown to be beneficial in XML retrieval, which is not surprising. Elements are of very different sizes, so it is important to have this incorporated when decided their relevance. Also elements are much smaller than whole documents, so again it makes sense to use additional evidence to estimate an element relevance. It is important to look at other types of evidence. It is also important to relate the evidence being combined, the combination framework (or the ranking strategy) and the indexing strategy. It is not yet clear how these influence each other, if at all. This investigation requires a (meta) formalism where all three can be modeled, reasoned about, and compared. Possible formalisms for carrying out this investigation include the matrix framework by Roelleke et al. [155] and region algebras by Mihajlovic et al. [129].

Different presentation paradigms are possible with XML retrieval, some of which have been investigated at INEX through dedicated retrieval tasks. This book touched little upon interface issues. The book mostly provided a description of presentation strategies for XML retrieval together with algorithms implementing these. It should, however, be stressed that presentation and interface issues are related. For example, presenting results from the relevant in context task requires an interface where users can easily zoom in and out a document. There has not been extensive work on interface issues for XML retrieval, but see the work of [122, 51, 11]. It has mostly been the case that interfaces were designed to study user behaviors in XML retrieval, and not to compare interfaces for XML retrieval. Because of the many ways to present results in XML retrieval, it is important to research into the design of appropriate interfaces for XML retrieval.

An important aspect of building an information retrieval system is its evaluation. Information retrieval research has a rich history in the development of methodology for evaluating how a retrieval system performs, i.e. how effective it is a returning the most relevant documents and only those. Approaches developed for retrieving XML elements need also to be evaluated with respect to their effectiveness. INEX developed the evaluation methodology to measure XML retrieval effectiveness. To consider that elements are of varying sizes and that there are structural relationships between elements, INEX had to opt for different ways to qualify relevance, to gather assessment data, and to measure effectiveness, compared to what is conventionally done in "flat" information retrieval. After a number of trials and errors, INEX has reached a stable stage, where it is now possible to investigate the effect of query languages, indexing strategies, ranking algorithms, and presentation paradigms on XML retrieval effectiveness. It would be interesting to see how the evaluation methodology developed by INEX, including the lessons learned, can benefit new (and more complex) information retrieval tasks, for example, in social networks, collaborative search, and aggregated search.

XML retrieval aims to provide focused access to documents marked up in XML by exploiting the XML mark-up. Although lots of advance has been made, the use of content-and-structure queries has not led to any significant increase of retrieval performance. In other words, there seems to be little benefit in expressing structural constraints in XML retrieval. It could be concluded that techniques to process structural constraints to provide a ranked list of result elements have not been effective. This is likely to be one of the reasons. There are others.

The structural constraints were to be interpreted as hints as where to find relevant content. As a consequence, relevance in INEX has been assessed with respect to content only. It is thus not surprising that structural constraints in queries have not helped in better identifying relevant elements. However, even if the structure was considered during assessments, it is still unlikely that the use of content-and-structure queries, as opposed to content-only queries, would yield any significant and consistent increase in retrieval performance.

While a main benefit of exploiting the XML mark-up is the semantics associated with each tag, the tags in the collections used at INEX (both the IEEE and the Wikipedia) have limited semantics. The XML mark-up in these collections specify a very basic "discourse" structure of a document, its title, abstract, sections and paragraphs, and so on. As a consequence, as for instance shown by Trotman et al. [181], the topics used at INEX are mainly targeting full documents, sections or paragraphs. Targeting document is document retrieval, targeting paragraph is passage retrieval, and targeting section is specifying a result size. The INEX topics are not about targeting some specific structural elements, which is why XML "element" retrieval has not (yet) benefited from content-and-structure queries.

It is important that INEX provides collections for which the XML mark-up goes beyond basic document discourse, in order to show the added values of content-and-structure queries for XML retrieval. INEX 2009 is making a step towards this by building a test collection with a much richer set of tags.

9.2 BEYOND XML "ELEMENT" RETRIEVAL

This book described the main components of an XML "element" retrieval system, including its evaluation. XML retrieval has other research challenges. Many of them have been or are being investigated as separate tracks at INEX, which we briefly introduce in this section. More details can be found on the INEX web site, and the various reports published in the SIGIR forum[1]. A chronology of the tracks at INEX can be found in [89].

Although text is dominantly present in most XML document collections, other types of media can also be found in those collections. Existing research on multimedia information retrieval has already shown that it is far from trivial to determine the combined relevance of a document that contains several multimedia objects. The objective of the **multimedia track** at INEX was to exploit the XML structure that provides a logical level at which multimedia objects are connected, to improve the retrieval performance of an XML-driven multimedia information retrieval system. One task, for instance, was to retrieve relevant document fragments based on an information need with a (structured) multimedia character [200, 191].

An important case scenario in, for instance, digital libraries is the retrieval of XML documents coming from heterogeneous sources. The **heterogeneous track** at INEX focused on the construction of an appropriate test collection and of appropriate tools for the evaluation of heterogeneous retrieval. It also investigates effective ways to retrieve from heterogeneous collections of XML documents [169, 45]. When building and querying heterogeneous XML repositories, a key problem is learning automatically from XML collections the relations between different formats and the transformations between different structured document representations (e.g. XML Scheme). This is currently been looked into by the **document mining track** at INEX, which aim is to explore algorithmic, theoretical and practical issues regarding the classification, clustering and structure mapping of structured information [41, 42].

[1] http://www.sigir.org/forum/index.html

The Wikipedia documents, upon which the INEX test collection is based since 2006, are extensively hyperlinked. The **Link-the-Wiki track** aimed to create a reusable resource for evaluating and comparing different state of the art systems and approaches to automated link discovery. More specifically, given a new orphan Wikipedia document, the aim is to recommend a set of incoming and outgoing links from/to anchor text in the existing collection. In the context of INEX, the aim is to operate at the XML element level. This means that anchor text or anchor elements linked not only to a related document, but also to a specific XML element within, or to the best entry point for starting to read the referenced material from [72]. The latter complements nicely the best in context retrieval task investigated at INEX.

The expert search task at the Enterprise track at TREC has evaluated systems that return a list of entities (people names) knowledgeable about a certain topic (e.g., "information retrieval") [16]. This can be generalized to arbitrary entity types. Consider, for example, the task of finding famous actors. Given a topic "1930s" it should return Astaire, Chaplin, and Gable, whereas given a topic "action" should result in Schwarzenegger, Stallone and Van Damme. The **entity ranking track** at INEX aimed to evaluate how well systems can rank entities in response to a user query, where the set of entities to be ranked is assumed to be loosely defined by a generic category, given in the query itself, or by some example entities [38].

The goal of an information retrieval system is not only to return relevant content to users, but to do so in a way that properly supports the interaction between users and the system. To study interaction in XML retrieval, INEX runs an **interactive track** dedicated to the investigation of the behavior of users when interacting with components of XML documents. Outcomes of the tracks were crucial in providing a better understanding of users (and their expectations) of XML retrieval systems. These outcomes influenced the main ad hoc retrieval track. For example, they led to the definition of retrieval tasks (focused retrieval task, relevance in context task, etc). Indeed, the presentation of results was found to be a very important aspect of XML retrieval, one that needed careful and separate investigation [177].

INEX devised, through its **use-cases track**, a thought experiment into how an XML retrieval will be used. Participants were asked to hypothesize how XML retrieval might be used in realistic situations. They were asked to hold local round-table discussions amongst themselves and with potential users, to participate in email discussions, and to present their findings in written form. The track provided a list of use cases able to illustrate how the main ad hoc retrieval tasks can be used in a realistic situation [140]. In addition, a taxonomy of use-cases was proposed [118], which has served as a basis for the refinement of retrieval tasks investigated INEX, including the set up of new tracks, e.g. the book search track.

Searching for information in a collection of books is indeed seen as one of the natural application areas of XML retrieval approaches, where a clear benefit to users is to gain direct access to parts of books relevant to their information need. The goal of the **book search track** is to investigate book-specific relevance ranking strategies, user interface issues and user behavior, exploiting special features, such as back of book indexes provided by authors, and linking to associated metadata like catalogue information from libraries. The track focuses on three themes, namely retrieval techniques for searching collections of digitized books, users' interactions with eBooks, and mechanisms to increase accessibility to the content of digitized books [90, 91].

An important issue in information retrieval is query formulation. In XML retrieval, query formulation has been investigated with respect to two scenarios. The first one consists of an alternative natural language expression of user information needs. Standard INEX topics with structural constraints are expressed in a query language (NEXI) that is not intended to be user-friendly. The main issue that was addressed by the **natural language query processing track** was to see whether it is practical to perform effective retrieval on the basis of a natural language expression that is the equivalent of a NEXI query. The outcomes of the track suggest that a natural language alternative to NEXI is indeed viable [61].

The second scenario was that of relevance feedback. In standard information retrieval, relevance feedback [156] has been translated into detecting a "bag of words" that are good (or bad) at retrieving relevant information. These terms are then added to (or removed from) the query and/or weighted according to their power in retrieving relevant information. With XML documents, a more sophisticated approach is needed. The **relevance feedback track** looked at how to exploit the characteristics of XML to infer which content and structural constraints lead to better query (re-)formulation [35, 36].

An issue not covered in this book is that of efficiency. Many of the retrieval strategies involve complex processing. Not only queries require more processing (e.g., due to their structural constraints), but the number of potentially retrievable elements is much larger than if dealing with whole-document only. In addition, the presentation of retrieval results (e.g. removing overlaps) requires additional processing. Many of the approaches evaluated during the INEX campaigns have mainly been concerned with effectiveness and less with efficiency. Some notable contributions, however, include [46, 173]. Since 2008, INEX has an **efficiency track**, which aims at providing a common forum for the evaluation of both the effectiveness and efficiency of XML retrieval approaches. The track is investigating different types of queries and retrieval scenarios, involving for instance queries with a deeply nested structure [175].

9.3 BEYOND XML RETRIEVAL

Information retrieval is not the only area of research that is concerned with providing effective and efficient information access. In the context of XML documents, another area of research is database. As the number of applications for accessing content- and data-oriented XML documents will increase (in digital libraries, intranet), access tools that cater for both types of documents will be increasingly needed. An important area of research in information access is the so-called integration of information retrieval and database technologies [6]. In the content of XML retrieval, research in information retrieval and databases have been concerned with different aspects of the information access process, e.g. ranking in information retrieval versus efficiency in databases. There is, however, a convergence between these two, as evidenced in a recent workshop on "Ranking XML query", where the latest results and challenges regarding this integration in the context of XML retrieval were exchanged [8]. [162] provides a description of this research area in the context of XML retrieval.

XML retrieval is one instance of focused retrieval. We recall that the latter is concerned with returning the most focused results to a given query. Other instances of focused retrieval include passage retrieval, and question & answer system. In principle, a query (or variants of a query) can be submitted to several focused retrieval systems if deemed appropriate. The results from each system would then need to be combined to form a so-called "aggregated result page". The aggregation of focused results is a new challenge in information access. For instance, a call for papers for a journal special issue on focussed retrieval and aggregated results was distributed in summer 2008. One topic of this special issue is the natural extension of focused retrieval, and in particular XML retrieval, to that of result aggregation. The task of aggregating results from various sources into one result page is an active research topic, and has been referred to as aggregated search [110]. Examples of aggregated search for the web include Yahoo! Alpha, Google Universal Search, and the Korean search engine Naver. It will be interesting to see how research on the aggregation of focused results and web aggregated search develops, including how they will influence each other. For instance, the evaluation of aggregated search can benefit from INEX, as both XML retrieval and aggregated search share the challenging task of how to present results to users.

Bibliography

[1] Agosti, M. Hypertext and information retrieval. *Information Processing & Management 29*, 3 (1993), 283–286. DOI: 10.1016/0306-4573(93)90055-I

[2] Ali, S., Consens, M., Kazai, G., and Lalmas, M. Structural relevance: a common basis for the evaluation of structured document retrieval, In *ACM CIKM International Conference on Information and Knowledge Management, Napa Valley, US* (2008) pp. 1153–1162.

[3] Alink, W., Bhoedjang, R., Boncz, P., and de Vries, A. XIRAF - XML-based indexing and querying for digital forensics. *Digital Investigation 3*, Supplement-1 (2006), 50–58. DOI: 10.1016/j.diin.2006.06.016

[4] Amati, G., Carpineto, C., and Romano, G. Merging XML Indices. In *Advances in XML Information Retrieval, Third International Workshop of the Initiative for the Evaluation of XML Retrieval, INEX 2004, Dagstuhl Castle, Germany, Revised Selected Papers* (2005), pp. 253–260. DOI: 10.1007/b136250

[5] Amer-Yahia, S., Botev, C., Dörre, J., and Shanmugasundaram, J. Full-Text extensions explained. *IBM Systems Journal 45*, 2 (2006), 335–352.

[6] Amer-Yahia, S., Case, P., Roelleke, T., Shanmugasundaram, J., and Weikum, G. Report on the DB/IR panel at SIGMOD 2005. *SIGMOD Record 34*, 4 (2005), 71–74. DOI: 10.1145/1107499.1107514

[7] Amer-Yahia, S., Cho, S., and Srivastava, D. Tree Pattern Relaxation. In *Advances in Database Technology - EDBT 2002, 8th International Conference on Extending Database Technology, Prague, Czech Republic* (2002), pp. 496–513. DOI: 10.1007/3-540-45876-X_32

[8] Amer-Yahia, S., Hiemstra, D., Roelleke, T., Srivastava, D., and Weikum, G. Ranked XML Querying. In *Workshop on Ranked XML Querying* (Dagstuhl, Germany, 2008), S. Amer-Yahia, D. Srivastava, and G. Weikum, Eds., no. 08111 in Dagstuhl Seminar Proceedings.

[9] Amer-Yahia, S., and Lalmas, M. XML search: languages, INEX and scoring. *SIGMOD Record 35*, 4 (2006), 16–23. DOI: 10.1145/1228268.1228271

[10] Arvola, P., Junkkari, M., and Kekäläinen, J. Generalized contextualization method for XML information retrieval. In *ACM CIKM International Conference on Information and Knowledge Management, Bremen, Germany* (2005), pp. 20–27. DOI: 10.1145/1099554.1099561

[11] Arvola, P., Junkkari, M., and Kekalainen, J. Applying XML retrieval methods for result document navigation in small screen devices. In *Mobile and ubiquitous information access (MUIA 2006) at MobileHCI 2006* (2006), pp. 6–10.

[12] Ashoori, E. *Using Topic shifts in Content-oriented XML Retrieval.* PhD thesis, Queen Mary, University of London, 2009.

[13] Ashoori, E., Lalmas, M., and Tsikrika, T. Examining topic shifts in content-oriented XML retrieval. *International Journal on Digital Libraries 8*, 1 (2007), 39–60. DOI: 10.1007/s00799-007-0026-5

[14] Baeza-Yates, R. A., Maarek, Y. S., Rölleke, T., and de Vries, A. Third edition of the XML and information retrieval workshop and first workshop on integration of IR and DB (WIRD) jointly held at SIGIR 2004. *SIGIR Forum 38*, 2 (2004), 24–30. DOI: 10.1145/1041394.1041400

[15] Baeza-Yates, R. A., and Ribeiro-Neto, B. *Modern Information Retrieval.* ACM Press/Addison-Wesley, 1999.

[16] Bailey, P., de Vries, A. P., Craswell, N., and Soboroff, I. Overview of the TREC 2007 Enterprise Track. In *Sixteenth Text REtrieval Conference, TREC 2007* (2007).

[17] Bates, M. J., and Maack, M., Eds. *Encyclopedia of Library and Information Sciences.* Taylor & Francis Group, 2009. To Appear.

[18] Baumgarten, C. A probabilistic model for distributed information retrieval. In *Proceedings of the 20th Annual International ACM SIGIR Conference on Research and Development in Information Retrieval, July 27-31, 1997, Philadelphia, PA, USA* (1997), pp. 258–266. DOI: 10.1145/258525.258585

[19] Ben-Aharon, Y., Cohen, S., Grumbach, Y., Kanza, Y., Mamou, J., Sagiv, Y., Sznajder, B., and Twito, E. Searching in an XML corpus using content and structure. In *INEX 2003 Proceedings* (2003), pp. 46–52.

[20] Blanken, H., Grabs, T., Schek, H.-J., Schenkel, R., and Weikum, G., Eds. *Intelligent Search on XML Data, Applications, Languages, Models, Implementations, and Benchmarks* (2003), vol. 2818, Springer.

[21] Bordogna, G., and Pasi, G. Personalised indexing and retrieval of heterogeneous structured documents. *Inf. Retr. 8*, 2 (2005), 301–318. DOI: 10.1007/s10791-005-5664-x

[22] Broschart, A., Schenkel, R., Theobald, M., and Weikum, G. TopX @ INEX 2007. In *Focused access to XML documents, 6th International Workshop of the Initiative for the Evaluation of XML Retrieval, INEX 2007, Dagstuhl Castle, Germany, Selected Papers* (2008). DOI: 10.1007/978-3-540-85902-4_4

[23] Burkowski, F. Retrieval activities in a database consisting of heterogeneous collections of structured text. In *15th Annual International ACM SIGIR Conference on Research and Development in Information Retrieval. Copenhagen, Denmark* (1992), pp. 112–125. DOI: 10.1145/133160.133185

[24] Callan, J. Passage-level evidence in document retrieval. In *Proceedings of the 17th annual international ACM SIGIR conference on Research and development in information retrieval* (1994), Springer-Verlag New York, Inc., pp. 302–310.

[25] Carmel, D., Maarek, Y., Mandelbrod, M., Y.Mass, and Soffer, A. Searching XML documents via XML fragments. In *26th Annual International ACM SIGIR Conference on Research and Development in Information Retrieval, Toronto, Canada* (2003), pp. 151–158. DOI: 10.1145/860435.860464

[26] Chamberlin, D., Robie, J., and Florescu, D. Quilt: An XML Query Language for Heterogeneous Data Sources. In *The World Wide Web and Databases, Third International Workshop WebDB 2000, Dallas, Texas, USA, Selected Papers* (2000), pp. 1–25.

[27] Chiaramella, Y., Mulhem, P., and Fourel, F. A model for multimedia information retrieval. Tech. rep., University of Glasgow, 1996.

[28] Chinenyanga, T. T., and Kushmerick, N. Expressive Retrieval from XML Documents. In *24th Annual International ACM SIGIR Conference on Research and Development in Information Retrieval, New Orleans, Louisiana* (2001), pp. 163–171. DOI: 10.1145/383952.383982

[29] Clarke, C. Controlling overlap in content-oriented XML retrieval. In *28th annual international ACM SIGIR conference on Research and development in information retrieval, Salvador, Brazil* (2005), pp. 314–321. DOI: 10.1145/1076034.1076089

[30] Clarke, C. Range results in XML retrieval. In *Proceedings of the INEX 2005 Workshop on Element Retrieval Methodology, Second Edition* (2005), pp. 4–5.

[31] Clarke, C. A., Cormack, G., and Burkowski, F. An algebra for structured text search and a framework for its implementation. *Computer Journal 38*, 1 (1995), 43–56. DOI: 10.1093/comjnl/38.1.43

[32] Cohen, S., Mamou, J., Kanza, Y., and Sagiv, Y. XSEarch: A Semantic Search Engine for XML. In *29th International Conference on Very Large Data Bases, Berlin, Germany* (2003), pp. 45–56.

[33] Consens, M., and Milo, T. Algebras for Querying Text Regions. In *Proceedings of the Fourteenth ACM SIGACT-SIGMOD-SIGART Symposium on Principles of Database Systems, San Jose, California* (1995), pp. 11–22. DOI: 10.1145/212433.212437

[34] Croft, W. B., and Schek, H.-J. Introduction to the special issue on database and information retrieval integration. *VLDB Journal 17*, 1 (2008), 1–3. DOI: 10.1007/s00778-007-0074-x

[35] Crouch, C. J. Relevance feedback at the INEX 2004 workshop. *SIGIR Forum 39*, 1 (2005), 41–42. DOI: 10.1145/1067268.1067282

[36] Crouch, C. J. Relevance feedback at INEX 2005. *SIGIR Forum 40*, 1 (2006), 58–59. DOI: 10.1145/1147197.1147208

[37] de Vries, A., Kazai, G., and Lalmas, M. Tolerance to irrelevance: A user-effort oriented evaluation of retrieval systems without predefined retrieval unit. In *RIAO 2004 Conference on Coupling approaches, coupling media and coupling languages for information retrieval, Vaucluse, France* (2004), pp. 463–473.

[38] de Vries, A. P., Vercoustre, A.-M., Thom, J. A., Craswell, N., and Lalmas, M. Overview of the inex 2007 entity ranking track. In *Focused Access to XML Documents, 6th International Workshop of the Initiative for the Evaluation of XML Retrieval, INEX 2007, Dagstuhl Castle, Germany* (2007), pp. 245–251. DOI: 10.1007/978-3-540-85902-4_22

[39] Delgado, A., and Baeza-Yates, R. An Analysis of Query Languages for XML. *UPGRADE. The European Online Magazine for the IT Professional, Special Issue on Information Retrieval and the Web III*, 3 (2002).

[40] Denoyer, L., and Gallinari, P. The Wikipedia XML Corpus. *SIGIR Forum 40*, 1 (2006), 64–69. DOI: 10.1145/1147197.1147210

[41] Denoyer, L., and Gallinari, P. Report on the XML mining track at INEX 2005 and INEX 2006: categorization and clustering of XML documents. *SIGIR Forum 41*, 1 (2007), 79–90. DOI: 10.1145/1273221.1273230

[42] Denoyer, L., and Gallinari, P. Report on the XML mining track at INEX 2007 categorization and clustering of XML documents. *SIGIR Forum 42*, 1 (2008), 22–28. DOI: 10.1145/1394251.1394255

[43] Deutsch, A., Fernández, M., Florescu, D., Levy, A., and Suciu, D. XML-QL. In *Query Languages 1998* (1998).

[44] Dopichaj, P. Element retrieval in digital libraries: Reality check. In *Proceedings of the SIGIR 2006 Workshop on XML Element Retrieval Methodology* (2006), pp. 1–4.

[45] Frommholz, I., and Larson, R. R. The Heterogeneous Collection Track at INEX 2006. In *Comparative Evaluation of XML Information Retrieval Systems, 5th International Workshop of the Initiative for the Evaluation of XML Retrieval, INEX 2006* (2006), pp. 312–317. DOI: 10.1145/1273221.1273229

[46] Fuhr, N., and Gövert, N. Retrieval quality vs. effectiveness of specificity-oriented search in XML collections. *Information Retrieval 9*, 1 (2006), 55–70. DOI: 10.1007/s10791-005-5721-5

[47] Fuhr, N., Gövert, N., Kazai, G., and Lalmas, M., Eds. *INitiative for the Evaluation of XML Retrieval (INEX). Proceedings of the First INEX Workshop. Dagstuhl, Germany, December 8–11, 2002* (Sophia Antipolis, France, 2003), ERCIM Workshop Proceedings, ERCIM.

[48] Fuhr, N., Gövert, N., and Rölleke, T. Dolores: A system for logic-based retrieval of multimedia objects. In *21st Annual International ACM SIGIR Conference on Research and Development in Information Retrieval, Melbourne, Australia* (1998), pp. 257–265. DOI: 10.1145/290941.291005

[49] Fuhr, N., and Großjohann, K. XIRQL: A Query Language for Information Retrieval in XML Documents. In *ACM SIGIR Conference on Research and Development in Information Retrieval, September 9-13, 2001, New Orleans, Louisiana, USA* (2001), pp. 172–180. DOI: 10.1145/383952.383985

[50] Fuhr, N., and Großjohann, K. XIRQL: An XML query language based on information retrieval concepts. *ACM Transaction on Information Systems 22*, 2 (2004), 313–356. DOI: 10.1145/984321.984326

[51] Fuhr, N., Großjohann, K., and Kriewel, S. A Query Language and User Interface for XML Information Retrieval. In *Intelligent Search on XML Data* (2003), pp. 59–75. DOI: 10.1007/b13249

[52] Fuhr, N., Kamps, J., Lalmas, M. ., Malik, S., and Trotman, A., Eds. *Focused Access to XML Documents, 6th International Workshop of the Initiative for the Evaluation of XML Retrieval, INEX 2007, Dagstuhl Castle, Germany, December 17-19, 2007. Selected Papers* (2008). DOI: 10.1007/978-3-540-85902-4

[53] Fuhr, N., Lalmas, M., and Malik, S., Eds. *INitiative for the Evaluation of XML Retrieval (INEX). Proceedings of the Second INEX Workshop. Dagstuhl, Germany, December 15–17, 2003* (2004).

[54] Fuhr, N., Lalmas, M., Malik, S., and Kazai, G., Eds. *Advances in XML Information Retrieval and Evaluation: Fourth Workshop of the INitiative for the Evaluation of XML Retrieval (INEX 2005)* (2006), vol. 3977 of *Lecture Notes in Computer Science*, Springer-Verlag.

[55] Fuhr, N., Lalmas, M., Malik, S., and Szlávik, Z., Eds. *Advances in XML Information Retrieval, Third International Workshop of the Initiative for the Evaluation of XML Retrieval, INEX 2004, Dagstuhl Castle, Germany, December 6-8, 2004, Revised Selected Papers* (2005), vol. 3493 of *Lecture Notes in Computer Science*, Springer.

[56] Fuhr, N., Lalmas, M., and Trotman, A., Eds. *Comparative Evaluation of XML Information Retrieval Systems, 5th International Workshop of the Initiative for the Evaluation of XML Retrieval, INEX 2006* (2007), vol. 4518 of *Lecture Notes in Computer Science*, Springer-Verlag.

[57] Gery, M., Largeron, C., and Thollard, F. Probabilistic document model integrating XML structure. In *INEX 2007 Pre-proceedings* (2007), pp. 139–149.

[58] Geva, S. GPX - Gardens Point XML IR at INEX 2005. In *Advances in XML Information Retrieval and Evaluation, 4th International Workshop of the Initiative for the Evaluation of XML Retrieval, INEX 2005, Dagstuhl Castle, Germany, Revised Selected Papers* (2006), pp. 240–253. DOI: 10.1007/11766278_18

[59] Geva, S. GPX - Gardens Point XML IR at INEX 2006. In *Comparative Evaluation of XML Information Retrieval Systems, 5th International Workshop of the Initiative for the Evaluation of XML Retrieval, INEX 2006* (2006), pp. 137–150. DOI: 10.1007/11766278_18

[60] Geva, S., Kamps, J., and Trotman, A., Eds. *INEX 2008 Workshop Pre-proceedings* (2008).

[61] Geva, S., and Woodley, A. The NLP task at INEX 2005. *SIGIR Forum 40*, 1 (2006), 60–63. DOI: 10.1145/1147197.1147209

[62] Goevert, N., Fuhr, N., Lalmas, M., and Kazai, G. Evaluating the effectiveness of content-oriented XML retrieval methods. *Journal of Information Retrieval 9*, 6 (2006), 699–722. DOI: 10.1007/s10791-006-9008-2

[63] Gonnet, G., and Tompa, F. W. Mind your grammar: a new approach to modelling text. In *13th International Conference on Very Large Data Bases, Brighton, England* (1987), pp. 339–346.

[64] Gövert, N., Abolhassani, M., N.Fuhr, and Großjohann, K. Content-oriented XML retrieval with HyRex. In *First Workshop of the INitiative for the Evaluation of XML Retrieval (INEX), Schloss Dagstuhl, Germany* (2002), pp. 26–32.

[65] Gövert, N., and Kazai, G. Overview of the INitiative for the Evaluation of XML retrieval (INEX) 2002. In *First Workshop of the INitiative for the Evaluation of XML Retrieval (INEX), Schloss Dagstuhl, Germany* (2002), pp. 1–17.

[66] Guo, L., Shao, F., Botev, C., and Shanmugasundaram, J. XRANK: Ranked Keyword Search over XML Documents. In *SIGMOD Conference* (2003), pp. 16–27.

[67] Harman, D. Overview of the First Text REtrieval Conference (TREC-1). In *TREC* (1992), pp. 1–20. DOI: 10.1016/0306-4573(93)90037-E

[68] Harman, D. The TREC conferences. In *Hypermedia - Information - Multimedia Conference* (1995), pp. 9–28.

[69] Hawking, D., Voorhees, E., N.Craswell, and Bailey, P. Overview of the TREC-8 Web Track. In *The Eighth Text REtrieval Conference (TREC-8), Gaithersburg, Maryland, National Institute of Standards and Technology (NIST)* (1999).

[70] Hearst, M. TextTiling: Segmenting Text into Multi-Paragraph Subtopic Passages. *Computational Linguistics 23*, 1 (1997), 33–64.

[71] Hiemstra, D., and Baeza-Yates, R. Structured text retrieval models. In *Encyclopedia of Database Systems*. Springer, 2009. To Appear.

[72] Huang, D. W. C., Xu, Y., Trotman, A., and Geva, S. Overview of INEX 2007 Link the Wiki Track. In *Focused Access to XML Documents, 6th International Workshop of the Initiative for the Evaluation of XML Retrieval, INEX 2007* (2007), pp. 373–387. DOI: 10.1007/978-3-540-85902-4_32

[73] Huang, F., Watt, S., Harper, D., and Clark, M. Compact representations in XML retrieval. In *Comparative Evaluation of XML Information Retrieval Systems, 5th International Workshop of the Initiative for the Evaluation of XML Retrieval, INEX 2006, Dagstuhl Castle, Germany, Revised and Selected Papers* (2006), pp. 64–72. DOI: 10.1007/978-3-540-73888-6_7

[74] Hubert, G. XML Retrieval Based on Direct Contribution of Query Components. In *Advances in XML Information Retrieval and Evaluation, 4th International Workshop of the Initiative for the Evaluation of XML Retrieval, INEX 2005, Dagstuhl Castle, Germany, 2005, Revised Selected Papers* (2006), pp. 172–186. DOI: 10.1007/11766278_13

[75] Itakura, K., and Clarke, C. A. University of waterloo at inex2007: Adhoc and link-the-wiki tracks. In *Focused Access to XML Documents, 6th International Workshop of the Initiative for the Evaluation of XML Retrieval, INEX 2007, Dagstuhl Castle, Germany, December 17-19, 2007. Selected Papers* (2007), pp. 417–425. DOI: 10.1007/978-3-540-85902-4_35

[76] Järvelin, K., and Kekäläinen, J. Cumulated gain-based evaluation of ir techniques. *ACM Transactions on Information Systems (ACM TOIS) 20*, 4 (2002), 422–446. DOI: 10.1145/582415.582418

[77] Joachims, T., Li, H., Liu, T.-Y., and Zhai, C. Learning to rank for information retrieval (LR4IR 2007). *SIGIR Forum 41*, 2 (2007), 58–62. DOI: 10.1145/1328964.1328974

[78] Kamps, J., de Rijke, M., and Sigurbjörnsson, B. Length normalization in XML retrieval. In *27th Annual International ACM SIGIR Conference on Research and Development in Information Retrieval, Sheffield, UK* (2004), pp. 80–87. DOI: 10.1145/1008992.1009009

[79] Kamps, J., Koolen, M., and Sigurbjörnsson, B. Filtering and clustering xml retrieval results. In *Comparative Evaluation of XML Information Retrieval Systems, 5th International Workshop of the Initiative for the Evaluation of XML Retrieval, INEX 2006* (2006), pp. 121–136. DOI: 10.1007/978-3-540-73888-6_13

[80] Kamps, J., Marx, M., de Rijke, M., and Sigurbjörnsson, B. Structured queries in XML retrieval. In *CIKM'05: Proceedings of the 14th ACM International Conference on Information and Knowledge Management* (2005), pp. 2–11. DOI: 10.1145/1099554.1099559

[81] Kamps, J., Pehcevski, J., Kazai, G., Lalmas, M., and Robertson, S. INEX 2007 Evaluation Metrics. In *Focused access to XML documents, 6th International Workshop of the Initiative for the Evaluation of XML Retrieval, INEX 2007, Dagstuhl Castle, Germany, Selected Papers* (2008). DOI: 10.1007/978-3-540-85902-4_2

[82] Kamps, J., and Sigurbjörnsson, B. What do users think of an XML element retrieval system? In *Advances in XML Information Retrieval and Evaluation: Fourth Workshop of the INitiative for the Evaluation of XML Retrieval (INEX 2005)* (2006), vol. 3977 of *Lecture Notes in Computer Science*, pp. 411–421.

[83] Kando, N., and Adachi, J. Report from the ntcir workshop 3. *SIGIR FORUM 38*, 1 (June 2004), 10–16. DOI: 10.1145/986278.986280

[84] Kapms, J. Indexing units. In *Encyclopedia of Database Systems*. Springer, 2009. To Appear.

[85] Kapms, J. Presenting structured text retrieval results. In *Encyclopedia of Database Systems*. Springer, 2009. To Appear.

[86] Kaszkiel, M., and Zobel, J. Passage retrieval revisited. In *Proceedings of the 20th annufal international ACM SIGIR conference on Research and development in information retrieval* (1997), ACM Press, pp. 178–185. DOI: 10.1145/258525.258561

[87] Kaszkiel, M., and Zobel, J. Effective ranking with arbitrary passages. *J. Am. Soc. Inf. Sci. Technol. 52*, 4 (2001), 344–364. DOI: 10.1002/1532-2890(2000)9999:9999<::AID-ASI1075>3.3.CO;2-R

[88] Kazai, G. Choosing an Ideal Recall-Base for the Evaluation of the Focused Task: Sensitivity Analysis of the XCG Evaluation Measures. In *Comparative Evaluation of XML Information Retrieval Systems, 5th International Workshop of the Initiative for the Evaluation of XML Retrieval, INEX 2006, Dagstuhl Castle, Germany, Revised and Selected Papers* (2007), pp. 35–44. DOI: 10.1007/978-3-540-73888-6_4

[89] Kazai, G. INitiative for the Evaluation of XML retrieval (INEX). In *Encyclopedia of Database Systems*. Springer, 2009. To Appear.

[90] Kazai, G., and Doucet, A. Overview of the INEX 2007 Book Search Track (BookSearch '07). *SIGIR Forum 42*, 1 (2008), 2–15. DOI: 10.1145/1394251.1394253

[91] Kazai, G., Doucet, A., and Landoni, M. Overview of the INEX 2008 book track. In *Advances in Focused Retrieval: 7th International Workshop of the Initiative for the Evaluation of XML Retrieval (INEX 2008)* (2009).

[92] Kazai, G., and Lalmas, M. Notes on What to Measure in INEX. In *Proceedings of the INEX 2005 Workshop on Element Retrieval Methodology* (Glasgow, July 2005).

[93] Kazai, G., and Lalmas, M. eXtended Cumulated Gain Measures for the Evaluation of Content-oriented XML Retrieval. *ACM Transactions on Information Systems 24*, 4 (2006), 503–542. DOI: 10.1145/1185877.1185883

[94] Kazai, G., Lalmas, M., and de Vries, A. The overlap problem in content-oriented XML retrieval evaluation. In *27th Annual International ACM SIGIR Conference on Research and Development in Information Retrieval, Sheffield, UK* (2004), pp. 72–79. DOI: 10.1145/1008992.1009008

[95] Kazai, G., Lalmas, M., and Reid, J. Construction of a test collection for the focussed retrieval of structured documents. In *Advances in Information Retrieval, 25th European Conference on IR Research, ECIR 2003, Pisa, Italy* (2003), pp. 88–103. DOI: 10.1007/3-540-36618-0_7

[96] Kazai, G., Lalmas, M., and Rölleke, T. A Model for the Representation and Focussed Retrieval of Structured Documents Based on Fuzzy Aggregation. In *String Processing and Information Retrieva, Chile* (2001), pp. 123–135.

[97] Kazai, G., Lalmas, M., and Rölleke, T. Focussed structured document retrieval. In *String Processing and Information Retrieval, Lisbon, Portugal* (2002), pp. 241–247. DOI: 10.1007/3-540-45735-6_21

[98] Kazai, G., Masood, S., and Lalmas, M. A study of the assessment of relevance for the inex'02 test collection. In *Advances in Information Retrieval, 26th European Conference on IR Research, ECIR 2004, Sunderland, UK, April 5-7, 2004, Proceedings* (2004), pp. 296–310. DOI: 10.1007/b96895

[99] Kazai, G., and Rölleke, T. A Scalable Architecture for XML Retrieval. In *First Workshop of the INitiative for the Evaluation of XML Retrieval (INEX), Schloss Dagstuhl, Germany* (2002), pp. 49–56.

[100] Kekalainen, J., Arvola, P., and Junkkari, M. Contextualization. In *Encyclopedia of Database Systems*. Springer, 2009. To Appear.

[101] Kekäläinen, J., and Järvelin, K. Using graded relevance assessments in IR evaluation. *Journal of the American Society for Information Science and Technology 53*, 13 (2002), 1120–1129. DOI: 10.1002/asi.10137

[102] Kilpeläinen, P., and Mannila, H. Retrieval from hierarchical texts by partial patterns. In *Proceedings of the 16th Annual International ACM-SIGIR Conference on Research and Development in Information Retrieval. Pittsburgh, PA, USA, 1993* (1993), pp. 214–222. DOI: 10.1145/160688.160722

[103] Kimelfeld, B., Kovacs, E., Sagiv, Y., and Yahav, D. Using Language Models and the HITS Algorithm for XML Retrieval. In *Comparative Evaluation of XML Information Retrieval Systems, 5th International Workshop of the Initiative for the Evaluation of XML Retrieval, INEX 2006, Dagstuhl Castle, Germany, December 17-20, 2006, Revised and Selected Papers* (2006), pp. 253–260. DOI: 10.1007/978-3-540-73888-6_25

[104] Krause, P., and Clark, D. *Representing Uncertain Knowledge - An Artificial Intelligence Approach.* Intellect Books, 1993.

[105] Lafferty, J. D., and Zhai, C. Document language models, query models, and risk minimization for information retrieval. In *Proceedings of the 24th Annual International ACM SIGIR Conference on Research and Development in Information Retrieval (SIGIR-01)* (New York, USA, 2001), pp. 111–119. DOI: 10.1145/383952.383970

[106] Lalmas, M. A model for representing and retrieving heterogeneous structured documents based on evidential reasoning. *Computer Journal 42*, 7 (1999), 547–568. DOI: 10.1093/comjnl/42.7.547

[107] Lalmas, M., and Baeza-Yates, R. Structured Document Retrieval. In *Encyclopedia of Database Systems*. Springer, 2009. To Appear.

[108] Lalmas, M., Kazai, G., Kamps, J., Pehcevski, J., Piwowarski, B., and Robertson, S. Inex 2006 evaluation metrics. In Fuhr et al. [56].

[109] Lalmas, M., and Moutogianni, E. A dempster-shafer indexing for the focused retrieval of a hierarchically structured document space: Implementation and experiments on a web museum collection. In *Computer-Assisted Information Retrieval (Recherche d'Information et ses Applications) - RIAO 2000, 6th International Conference, College de France, France* (2000), pp. 442–456.

[110] Lalmas, M., and Murdock, V., Eds. *ACM SIGIR Workshop on Aggregated Search* (Singapore, 2008).

[111] Lalmas, M., and Rölleke, T. Modelling vague content and structure querying in xml retrieval with a probabilistic object-relational framework. In *6th International Conference on Flexible Query Answering Systems, FQAS 2004, Lyon, France* (2004), pp. 432–445.

[112] Lalmas, M., and Ruthven, I. Representing and retrieving structured documents using the dempster-shafer theory of evidence: Modelling and evaluation. *Journal of Documentation 54*, 5 (1998), 529–565. DOI: 10.1108/EUM0000000007180

[113] Lalmas, M., and Tombros, A. Evaluating XML Retrieval Effectiveness at INEX. *SIGIR Forum 41*, 1 (2007), 40–57. DOI: 10.1145/1273221.1273225

[114] Lalmas, M., and Tombros, A. INEX 2002 - 2006: Understanding XML Retrieval Evaluation. In *Digital Libraries: Research and Development, First International DELOS Conference, Pisa, Italy* (2007), pp. 187–196. DOI: 10.1007/978-3-540-77088-6_18

[115] Larsen, B., Tombros, A., and Malik, S. Obtrusiveness and relevance assessment in interactive XML IR experiments. In *Proceedings of the INEX 2005 Workshop on Element Retrieval Methodology* (2007), pp. 39–42.

[116] Larson, R. Cheshire II at INEX 03: Component and Algorithm Fusion for XML Retrieval. In *INEX 2003 Proceedingss* (2003), pp. 38–45.

[117] Larson, R. Probabilistic retrieval, component fusion and blind feedback for XML retrieval. In *Advances in XML Information Retrieval and Evaluation, 4th International Workshop of the Initiative for the Evaluation of XML Retrieval, INEX 2005, Dagstuhl Castle, Germany, Revised Selected Papers* (2006), pp. 225–239. DOI: 10.1007/11766278_17

[118] Lehtonen, M., Pharo, N., and Trotman, A. A Taxonomy for XML Retrieval Use Cases. In *Comparative Evaluation of XML Information Retrieval Systems, 5th International Workshop of the Initiative for the Evaluation of XML Retrieval, INEX 2006* (2006).

[119] Lu, W., Robertson, S., and MacFarlane, A. Field-Weighted XML Retrieval Based on BM25. In *Advances in XML Information Retrieval and Evaluation, 4th International Workshop of the Initiative for the Evaluation of XML Retrieval, INEX 2005, Dagstuhl Castle, Germany, Revised Selected Papers* (2005), pp. 161–171.

[120] MacLeod, I. A query language for retrieving information from hierarchic text structures. *The Computer Journal 34*, 3 (1991), 254–264. DOI: 10.1093/comjnl/34.3.254

[121] Malik, S., Kazai, G., Lalmas, M., and Fuhr, N. Overview of inex 2005. In *Advances in XML Information Retrieval and Evaluation, 4th International Workshop of the Initiative for the Evaluation of XML Retrieval, INEX 2005* (2005), pp. 1–15.

[122] Malik, S., Klas, C.-P., Fuhr, N., Larsen, B., and Tombros, A. Designing a User Interface for Interactive Retrieval of Structured Documents - Lessons Learned from the INEX Interactive Track. In *Research and Advanced Technology for Digital Libraries, 10th European Conference, ECDL 2006* (2006), pp. 291–302. DOI: 10.1007/11863878_25

[123] Malik, S., Trotman, A., Lalmas, M., and Fuhr, N. Overview of INEX 2006. In *Comparative Evaluation of XML Information Retrieval Systems, 5th International Workshop of the Initiative for the Evaluation of XML Retrieval, INEX 2006, Dagstuhl Castle, Germany, December 17-20, 2006, Revised and Selected Papers* (2007), pp. 1–11. DOI: 10.1007/978-3-540-73888-6_1

[124] Manning, C. D., Raghavan, P., and Schuetze, H. *Introduction to Information Retrieval.* Cambridge University Press, 2008.

[125] Mass, Y. IBM HRL at INEX 06. In *Comparative Evaluation of XML Information Retrieval Systems, 5th International Workshop of the Initiative for the Evaluation of XML Retrieval, INEX 2006* (2006), pp. 151–159.

[126] Mass, Y., and Mandelbrod, M. Retrieving the most relevant XML Components. In *INEX 2003 Proceedings* (2003), pp. 53–58.

[127] Mass, Y., and Mandelbrod, M. Component Ranking and Automatic Query Refinement for XML Retrieval. In *Advances in XML Information Retrieval, Third International Workshop of the Initiative for the Evaluation of XML Retrieval, INEX 2004, Dagstuhl Castle, Germany, Revised Selected Papers* (2005), pp. 73–84.

[128] Mass, Y., and Mandelbrod, M. Using the INEX Environment as a Test Bed for Various User Models for XML Retrieval. In *Advances in XML Information Retrieval and Evaluation, 4th International Workshop of the Initiative for the Evaluation of XML Retrieval, INEX 2005, Dagstuhl Castle, Germany, Revised Selected Papers.* (2006), pp. 187–195. DOI: 10.1007/11766278_14

[129] Mihajlovic, V., Blok, H. E., Hiemstra, D., and Apers, P. M. G. Score region algebra: building a transparent XML-IR database. In *ACM CIKM International Conference on Information and Knowledge Management, Bremen, Germany* (2005), pp. 12–19.

[130] Mihajlovic, V., Ramírez, G., Westerveld, T., Hiemstra, D., Blok, H. E., and de Vries, A. TIJAH Scratches INEX 2005: Vague Element Selection, Image Search, Overlap, and Relevance Feedback. In *Advances in XML Information Retrieval and Evaluation, 4th International Workshop of the Initiative for the Evaluation of XML Retrieval, INEX 2005, Dagstuhl Castle, Germany, Revised Selected Papers* (2006), pp. 72–87. DOI: 10.1007/11766278_6

[131] Mittendorf, E., and Schäuble, P. Document and passage retrieval based on hidden markov models. In *Proceedings of the 17th annual international ACM SIGIR conference on Research and development in information retrieval* (1994), Springer-Verlag New York, Inc., pp. 318–327.

[132] Myaeng, S.-H., Jang, D.-H., Kim, M.-S., and Zhoo, Z.-C. A flexible model for retrieval of SGML documents. In *21st Annual International ACM SIGIR Conference on Research and Development in Information Retrieval, Melbourne, Australia* (1998), pp. 138–145. DOI: 10.1145/290941.290980

[133] Navarro, G., and Baeza-Yates, R. Proximal nodes: A model to query document databases by content and structure. *ACM Transactions on Information Systems 15*, 4 (1997), 400–435. DOI: 10.1145/263479.263482

[134] Ogilvie, P., and Lalmas, M. Investigating the exhaustivity dimension in content-oriented XML element retrieval evaluation. In *ACM CIKM International Conference on Information and Knowledge Management, Arlington, Virginia, USA* (2006), pp. 84–93. DOI: 10.1145/1183614.1183631

[135] O'Keefe, R. A. If inex is the answer, what is the question? In *INEX* (2004), pp. 54–59.

[136] O'Keefe, R. A., and Trotman, A. The simplest query language that could possibly work. In *INEX 2003 Proceedings* (2003), pp. 167–174.

[137] Ozsu, M., and Liu, L., Eds. *Encyclopedia of Database Systems*. Springer, 2009. To Appear.

[138] Pehcevski, J., and Thom, J. A. Hixeval: Highlighting XML retrieval evaluation. In *Advances in XML Information Retrieval and Evaluation, 4th International Workshop of the Initiative for the Evaluation of XML Retrieval, INEX 2005, Dagstuhl Castle, Germany, Revised Selected Papers* (2006), pp. 43–57. DOI: 10.1007/11766278_4

[139] Peters, C., Braschler, M., Gonzalo, J., and Kluck, M., Eds. *Evaluation of Cross-Language Information Retrieval Systems (CLEF 2001)* (2002), vol. 2406, Springer. DOI: 10.1007/3-540-45691-0

[140] Pharo, N., and Trotman, A. The use case track at INEX 2006. *SIGIR Forum 41*, 1 (2007), 64–66. DOI: 10.1145/1273221.1273227

[141] Pinel-Sauvagnat, K. Propagation-based structured text retrieval. In *Encyclopedia of Database Systems*. Springer, 2009. To Appear.

[142] Piwowarski, B., and Dupret, G. Evaluation in (XML) information retrieval: expected precision-recall with user modelling (eprum). In *29th Annual International ACM SIGIR Conference on Research and Development in Information Retrieval, Seattle, Washington, USA* (2006), pp. 260–267.

[143] Piwowarski, B., and Lalmas, M. Providing consistent and exhaustive relevance assessments for XML retrieval evaluation. In *12th ACM international conference on Information and knowledge management, Washington, DC, USA* (2004), pp. 361–370.

[144] Piwowarski, B., Trotman, A., and Lalmas, M. Sound and complete relevance assessments for XML retrieval. *ACM Transactions in Information Systems 27*, 1 (2009). To Appear.

[145] P.Ogilvie, and Callan, J. Hierarchical language models for XML component retrieval. In *Advances in XML Information Retrieval, Third International Workshop of the Initiative for the Evaluation of XML Retrieval, INEX 2004, Dagstuhl Castle, Germany, Revised Selected Papers* (2005), pp. 224–237.

[146] Ponte, J. M., and Croft, W. B. A language modeling approach to information retrieval. In *SIGIR '98: Proceedings of the 21st Annual International ACM SIGIR Conference on Research and Development in Information Retrieval, August 24-28 1998, Melbourne, Australia* (1998), pp. 275–281. DOI: 10.1145/290941.291008

[147] Popovici, E., Ménier, G., and Marteau, P.-F. SIRIUS XML IR System at INEX 2006: Approximate Matching of Structure and Textual Content. In *Comparative Evaluation of XML Information Retrieval Systems, 5th International Workshop of the Initiative for the Evaluation of XML Retrieval, INEX 2006, Dagstuhl Castle, Germany, Revised and Selected Papers* (2007), pp. 185–199. DOI: 10.1007/978-3-540-73888-6_19

[148] Raghavan, V., Jung, G., and Bollmann, P. A critical investigation of recall and precision as measures of retrieval system performance. *ACM Transaction on Information Systems 7*, 3 (1989), 205–229. DOI: 10.1145/65943.65945

[149] Ramírez, G. *Structural Features in XML Retrieval.* PhD thesis, University of Amsterdam, 2007.

[150] Ramirez, G. Processing overlaps. In *Encyclopedia of Database Systems*. Springer, 2009. To Appear.

[151] Reid, J., Lalmas, M., Finesilver, K., and Hertzum, M. Best entry points for structured document retrieval - Part I: Characteristics. *Information Processing & Management 42*, 1 (2006), 74–88. DOI: 10.1016/j.ipm.2005.03.006

[152] Robertson, S., Zaragoza, H., and Taylor, M. Simple BM25 extension to multiple weighted fields. In *ACM CIKM International Conference on Information and Knowledge Management, Washington, DC, USA* (2004), pp. 42–49.

[153] Rölleke, T., and Fuhr, N. Retrieval of Complex Objects Using a Four-Valued Logic. In *9th Annual International ACM SIGIR Conference on Research and Development in Information Retrieval* (1996), pp. 206–214.

[154] Rölleke, T., Lalmas, M., Kazai, G., Ruthven, I., and Quicker, S. The accessibility dimension for structured document retrieval. In *Advances in Information Retrieval, 24th BCS-IRSG European Colloquium on IR Research, Glasgow, UK* (2002), pp. 284–302. DOI: 10.1007/3-540-45886-7_19

[155] Rölleke, T., Tsikrika, T., and Kazai, G. A general matrix framework for modelling information retrieval. *Inf. Process. Manage. 42*, 1 (2006), 4–30. DOI: 10.1016/j.ipm.2004.11.006

[156] Ruthven, I., and Lalmas, M. A survey on the use of relevance feedback for information access systems. *Knowledge Engineering Review 18*, 1 (2003).

[157] Salton, G., Ed. *The SMART Retrieval System - Experiments in Automatic Document Processing.* Prentice Hall, Englewood, Cliffs, New Jersey, 1971.

[158] Salton, G., Singhal, A., Buckley, C., and Mitra, M. Automatic text decomposition using text segments and text themes. In *Proceedings of the the seventh ACM conference on Hypertext* (1996), pp. 53–65. DOI: 10.1145/234828.234834

[159] Sauvagnat, K., Boughanem, M., and Chrisment, C. Answering content and structure-based queries on XML documents using relevance propagation. *Information Systems 31*, 7 (2006), 621–635. DOI: 10.1016/j.is.2005.11.007

[160] Sauvagnat, K., Hlaoua, L., and Boughanem, M. XFIRM at INEX 2005: Ad-Hoc and Relevance Feedback Tracks. In *Advances in XML Information Retrieval and Evaluation, 4th International Workshop of the Initiative for the Evaluation of XML Retrieval, INEX 2005, Dagstuhl Castle, Germany, Revised Selected Papers* (2006), pp. 88–103. DOI: 10.1007/11766278_7

[161] Schenkel, R., and Theobald, M. Structural Feedback for Keyword-Based XML Retrieval. In *Advances in Information Retrieval, 28th European Conference on IR Research, ECIR 2006, London, UK* (2006), pp. 326–337. DOI: 10.1007/11735106_29

[162] Schenkel, R., and Theobald, M. Integrated DB&IR. In *Encyclopedia of Database Systems*. Springer, 2009. To Appear.

[163] Schlieder, T., and Meuss, H. Querying and ranking XML documents. *JASIST 53*, 6 (2002), 489–503. DOI: 10.1002/asi.10060

[164] Sigurbjörnsson, B., and Kamps, J. The Effect of Structured Queries and Selective Indexing on XML Retrieval. In *Advances in XML Information Retrieval and Evaluation, 4th International Workshop of the Initiative for the Evaluation of XML Retrieval, INEX 2005* (2005), pp. 104–118.

[165] Sigurbjornsson, B., Kamps, J., and de Rijke, M. An Element-Based Approch to XML Retrieval. In *Proceedings INEX 2003 Workshop* (2004), pp. 19–26.

[166] Sparck-Jones, K. What's the value of TREC: is there a gap to jump or a chasm to bridge? *SIGIR Forum 40*, 1 (2006), 10–20.

[167] Sparck-Jones, K., and van Rijsbergen, C. J. Information retrieval test collections. *Journal of Documentation 32*, 1 (Aug. 1976), 59–75.

[168] Szlavik, Z. *Content and structure summarisation for accessing XML documents*. PhD thesis, Queen Mary, University of London, 2008.

[169] Szlávik, Z., and Rölleke, T. Building and experimenting with a heterogeneous collection. In *Advances in XML Information Retrieval, Third International Workshop of the Initiative for the Evaluation of XML Retrieval, INEX 2004* (2004), pp. 349–357.

[170] Szlávik, Z., Tombros, A., and Lalmas, M. The use of summaries in xml retrieval. In *Research and Advanced Technology for Digital Libraries, 10th European Conference, ECDL 2006, Alicante, Spain* (2006), pp. 75–86. DOI: 10.1007/11863878_7

[171] Szlávik, Z., Tombros, A., and Lalmas, M. Feature- and query-based table of contents generation for xml documents. In *Advances in Information Retrieval, 29th European Conference on IR Research, ECIR 2007, Rome, Italy, April 2-5, 2007, Proceedings* (2007), pp. 456–467. DOI: 10.1007/978-3-540-71496-5_41

[172] Theobald, A., and Weikum, G. The Index-Based XXL Search Engine for Querying XML Data with Relevance Ranking. In *EDBT* (2002), pp. 477–495.

[173] Theobald, M., Bast, H., Majumdar, D., R., and Weikum, G. TopX: efficient and versatile top-*k* query processing for semistructured data. *VLDB Journal 17*, 1 (2008), 81–115.

[174] Theobald, M., Schenkel, R., and Weikum, G. TopX and XXL at INEX 2005. In *Advances in XML Information Retrieval and Evaluation, 4th International Workshop of the Initiative for the Evaluation of XML Retrieval, INEX 2005, Dagstuhl Castle, Germany, Revised Selected Papers* (2006), pp. 282–295. DOI: 10.1007/11766278_21

[175] Theobald, M., and Shenkel, R. Overview of the INEX 2008 Efficiency track. In *Advances in Focused Retrieval: 7th International Workshop of the Initiative for the Evaluation of XML Retrieval (INEX 2008)* (2009).

[176] Tombros, A., Larsen, B., and Malik, S. The interactive track at INEX 2004. In *Advances in XML Information Retrieval, Third International Workshop of the Initiative for the Evaluation of XML Retrieval, INEX 2004, Dagstuhl Castle, Germany, Revised Selected Papers* (2005), pp. 410–423.

[177] Tombros, A., Malik, S., and Larsen, B. Report on the INEX 2004 interactive track. *SIGIR Forum 39*, 1 (2005), 43–49. DOI: 10.1145/1067268.1067283

[178] Trotman, A. Wanted: Element retrieval users. In *Proceedings of the INEX 2005 Workshop on Element Retrieval Methodology* (Glasgow, July 2005).

[179] Trotman, A. Narrowed Extended XPath I (NEXI). In *Encyclopedia of Database Systems*. Springer, 2009. To Appear.

[180] Trotman, A. Processing structural constraints. In *Encyclopedia of Database Systems*. Springer, 2009. To Appear.

[181] Trotman, A., del Rocio Gomez Crisostomo, M., and Lalmas, M. Visualizing the Problems with the INEX Topics. In *32nd Annual International ACM SIGIR Conference on Research and Development in Information Retrieval, Boston, USA* (2009).

[182] Trotman, A., Geva, S., and Kamps, J. Report on the SIGIR 2007 workshop on focused retrieval. *SIGIR Forum 41*, 2 (2007), 97–103. DOI: 10.1145/1328964.1328981

[183] Trotman, A., Kamps, J., and Geva, S., Eds. *Proceedings of the SIGIR 2008 Workshop on Focused Retrieval* (2008), University of Otago, Dunedin New Zealand.

[184] Trotman, A., and Lalmas, M. Strict and vague interpretation of XML-retrieval queries. In *29th Annual International ACM SIGIR Conference on Research and Development in Information Retrieval, Seattle, Washington, USA* (2006), pp. 709–710. DOI: 10.1145/1148170.1148329

[185] Trotman, A., and Lalmas, M. Why structural hints in queries do not help XML retrieval. In *29th Annual International ACM SIGIR Conference on Research and Development in Information Retrieval, Seattle, Washington, USA* (2006), pp. 711–712. DOI: 10.1145/1148170.1148330

[186] Trotman, A., and Sigurbjornsson, B. Narrowed Extended XPath I (NEXI). In *Advances in XML Information Retrieval, Third International Workshop of the Initiative for the Evaluation of XML Retrieval, INEX 2004, Dagstuhl Castle, Germany, Revised Selected Papers* (2005), pp. 16–40.

[187] Tsikrika, T. Aggregation-based Structured Text Retrieval. In *Encyclopedia of Database Systems*. Springer, 2009. To Appear.

[188] van Rijsbergen, C. J. *Information Retrieval*. Butterworth, 1979.

[189] van Rijsbergen, C. J., and Sparck-Jones, K. A test for the separation of relevant and non-relevant documents in experimental retrieval collections. *Journal of Documentation 29* (1973), 251–257. DOI: 10.1108/eb026557

[190] van Zwol, R. B^3-SDR and Effective Use of Structural Hints. In *Advances in XML Information Retrieval and Evaluation, 4th International Workshop of the Initiative for the Evaluation of XML Retrieval, INEX 2005, Dagstuhl Castle, Germany, Revised Selected Papers* (2006), pp. 146–160.

[191] van Zwol, R., Kazai, G., and Lalmas, M. INEX 2005 Multimedia Track. In *Advances in XML Information Retrieval and Evaluation, 4th International Workshop of the Initiative for the Evaluation of XML Retrieval, INEX 2005* (2005), pp. 497–510.

[192] Vittaut, J.-N., and Gallinari, P. Machine Learning Ranking for Structured Information Retrieval. In *Advances in Information Retrieval, 28th European Conference on IR Research, ECIR 2006, London, UK, 2006,* (2006), pp. 338–349. DOI: 10.1007/11735106_30

[193] Vittaut, J.-N., Piwowarski, B., and Gallinari, P. An Algebra for Structured Queries in Bayesian Networks. In *Advances in XML Information Retrieval, Third International Workshop of the Initiative for the Evaluation of XML Retrieval, INEX 2004, Dagstuhl Castle, Germany, December 6-8, 2004, Revised Selected Papers* (2004), pp. 100–112.

[194] Voorhees, E., Gupta, N. K., and Johnson-Laird, B. Learning collection fusion strategies. In *18th Annual International ACM SIGIR Conference on Research and Development in Information Retrieval, Seattle, Washington, USA* (1995), pp. 172–179. DOI: 10.1145/215206.215357

[195] Voorhees, E. M. Overview of the trec 2002 question answering track. In *TREC* (2002).

[196] Voorhees, E. M., and Harman, D. K., Eds. *The Tenth Text REtrieval Conference (TREC 2001)* (Gaithersburg, MD, USA, 2002), NIST.

[197] Voorhees, E. M., and K., D. *TREC - Experiment and Evaluation in Information Retrieval.* MIT Press, 2005.

[198] Voorhess, E. TREC: Improving Information Access through Evaluation. *ASIS&T Bulletin 32*, 1 (2005).

[199] Westerveld, T., Rode, H., van Os, R., Hiemstra, D., Ramírez, G., Mihajlovic, V., and de Vries, A. P. Evaluating Structured Information Retrieval and Multimedia Retrieval Using PF/Tijah. In *Comparative Evaluation of XML Information Retrieval Systems, 5th International Workshop of the Initiative for the Evaluation of XML Retrieval, INEX 2006* (2006), pp. 104–114.

[200] Westerveld, T., and van Zwol, R. Multimedia retrieval at INEX 2006. *SIGIR Forum 41*, 1 (2007), 58–63. DOI: 10.1145/1273221.1273226

[201] Wilkinson, R. Effective retrieval of structured documents. In *Proceedings of the 17th annual international ACM SIGIR conference on Research and development in information retrieval* (1994), Springer-Verlag New York, Inc., pp. 311–317.

[202] Woodley, A., and Geva, S. NLPX - An XML-IR System with a Natural Language Interface. In *ADCS 2004, Proceedings of the Ninth Australasian Document Computing Symposium* (2004), pp. 71–74.

[203] Zwol, R., Baas, J., van Oostendorp, H., and Wiering, F. Bricks: The Building Blocks to Tackle Query Formulation in Structured Document Retrieval. In *Advances in Information Retrieval, 28th European Conference on IR Research, ECIR 2006, London* (2006), pp. 314–325. DOI: 10.1007/11735106_28

Biography

MOUNIA LALMAS

Professor Mounia Lalmas has taken up in September 2008 the position of Research Chair in Information Retrieval at the University of Glasgow funded by Microsoft Research/Royal Academy of Engineering. She was previously a Professor of Information Retrieval at Queen Mary, University of London, which she joined in 1999 as a lecturer. Prior to this, she was a Research Scientist at the University of Dortmund in 1998, a Lecturer from 1995 to 1997 and a Research Fellow from 1997 to 1998 at the University of Glasgow, where she received her PhD in 1996. Her research focuses on the development and evaluation of intelligent access to interactive heterogeneous and complex information repositories, and covering a wide range of domains such as Web, XML, and MPEG-7. From 2002 until 2007, she co-led with Norbert Fuhr the Evaluation Initiative for XML Retrieval (INEX), a large-scale project with over 80 participating organizations worldwide, which was responsible for defining the nature of XML retrieval, and how it should be evaluated. She has given numerous presentations and lectures on XML retrieval and evaluation, for instance at CIKM, SIGIR and ESSIR. She is now working on technologies for aggregated search and bridging the digital divide. She is also currently getting back into theoretical information retrieval where she is looking at the use of quantum theory to model interactive information retrieval. She is currently the ACM SIGIR Vice-Chair. She was the workshop co-chair at SIGIR 2004 and 2006, mentoring chair at SIGIR 2009, PR (co-) chair at CIKM 2008 and WI/IAT 2009, PC chair at ECIR 2006, vice co-chair for the XML and Web Data track at WWW 2009, and general co-chair of IIiX 2008. She serves on the TOIS, IP&M and IR journal editorial boards.